LIVING WITH LEAD

Intersections: Environment, Science, Technology

Sarah Elkind and Finn Arne Jørgensen, Editors

LIVING WITH LEAD

An Environmental History of Idaho's Coeur D'Alenes, 1885–2011

BRADLEY D. SNOW

UNIVERSITY OF PITTSBURGH PRESS

Published by the University of Pittsburgh Press, Pittsburgh, Pa., 15260

Manufactured in the United States of America
Printed on acid-free paper
10 9 8 7 6 5 4 3 2 1

Cataloging-in-Publication data is available from the Library of Congress

ISBN 13: 978-0-8229-6448-3
ISBN 10: 0-8229-6448-1

Cover design by Bruce Gore

To Reilly and the people of the Coeur d'Alenes

CONTENTS

ACKNOWLEDGMENTS

This book could not have been written and published without the help of numerous people. Friends, family, teacher-scholars, librarians, fellow graduate students, editors—all played a critical role in helping me to do the work you have before you. And much work it has been. This book has been nearly a decade in the making. I began doing archival research for the book in the late spring of 2007. Years of research were accompanied and followed by years of writing and rewriting. The original drafts of what became my doctoral dissertation have been rewritten at least a score of times.

From the Department of History, Philosophy and Religious Studies at Montana State University–Bozeman I received an excellent graduate education. My dissertation committee—Mary Murphy, Tim LeCain, Rob Campbell, Michael Reidy, and Dale Martin—provided great support and guidance, along with much-needed critique and occasional prodding, which vastly improved the quality of this work. Dale and Mary deserve the credit and/or blame for steering me to the eminently "heavy subject" of lead and the Coeur d'Alenes that since has become one of my chief preoccupations.

Mary Murphy, my committee chair, trained her keen eye for social history and bad writing on the project from its inception. She read and edited every line of every draft of the work through its publication as a dissertation, cutting reams of unnecessary verbiage, and greatly enhancing the quality of my prose. I owe Mary a great debt. Almost unfailingly upbeat, endlessly patient, relentlessly professional, Mary kept me going throughout the long years I spent "dissertating." Tim LeCain also read through multiple drafts of my dissertation and brought to my attention any number of important and relevant concepts and scholarly works from environmental and technological history. Michael Reidy did a line-by-line edit, including pointed margin notations, of a late version of the dissertation that did much to improve the work. Rob Campbell provided forceful and pointed critique of the dissertation that—much as it pained me at the time—pushed me to make the work more broadly applicable and theoretically stronger. Dale Martin offered a willing and highly knowledgeable conversant over all issues related to lead or the Silver Valley. Dale, a true polymath, kept me honest when it came to all of my mining, milling, and smelting facts and terminology and also drew the fine map of the Kellogg area for the book.

I would also like to thank the department at Montana State for allowing me to teach courses, affording me numerous teaching assistantships, and also for providing support that allowed me to attend two conferences of the American Society for Environmental History during my tenure as a doctoral candidate. Special thanks go to Bob Rydell, Brett Walker, and Tim LeCain for offering me other meaningful employment opportunities while I was a PhD candidate.

A doctoral program, of course, is more than just solitary work, courses, and meetings with professors. It is also a community of people. I was fortunate during my time at Montana State to be surrounded by a very fun, dynamic, and supportive group of fellow grad students. These include Mike Wise, Megan Raby, Jaime Allison, Paul McCutcheon, Andrew Johnson, Paul Sivitz, Dan Zizzamia, Matt Fockler (a geographer, but an honorary "history person" nonetheless), Bob Gardner, and Jerry Jessee. You all now have gone on to bigger and better things, but it was great to have had your love and support during those critical graduate school years. Matt, Jerry, Paul, Bob, and Andrew, thanks for critiquing an early version of the "Lead Creek" chapter of the book. Matt, thanks for helping me plan my trip to the US Forest Service Archives in Seattle. I would've struggled to find my way around there without your help. Jerry, thanks for your friendship, for the many discussions about the "environmental body" and other aspects of environmental history, and for your advice on using thematic chapters that got my writing "unstuck."

I did much of my archival work at the University of Idaho Library's Special Collections. A number of fine librarians, including Nathan Bender, helped me there, but Julie Monroe truly stands out for her assistance to this project. Julie proved nearly indispensable in her ability and cheerful willingness to help me track down documents and pictures from the library's sprawling Bunker Hill Collection. She was a pleasure to work with and her town, Moscow, Idaho, a pleasure to visit.

Over the years I have made numerous research trips to the Silver Valley. The people I met there have been unfailingly open, friendly, and interested in the history of their fascinating part of the world. I thank them for their openness and their hospitality to a stranger.

From the time I began talking with her about this book project at the end of 2014, Sandy Crooms at the University of Pittsburgh Press has been a delight to work with. Straightforward, encouraging, wise and thoughtful, professional, she has given this newcomer to book publishing an excellent experience. The rest of the staff at the Press with whom I've had the chance to work have likewise been nothing but a pleasure.

I spent more hours working at coffeehouses on this project than I would care to calculate. Much of that time was spent at Chadz, a wonderfully comfortable and welcoming place in my hometown of Livingston, Montana. Chadz's proprietor, Chad Petrulis, was very gracious about allowing this rather bookish and intense patron to monopolize space and endless free coffee refills in his establishment over a period of years.

My father has been interested in and supportive of all my efforts on this project from the start. Thanks, Dad.

Thanks also to my grandmother, Charlotte S. Wheeler, long since departed, who helped nurture in me a deep interest in history at an early age.

A project like this necessarily disrupts the lives of nuclear family members. I want to thank my wife, Reilly; my children, Nathan, Francesca, Grant, and Willa; and my stepson, Caen, for their forbearance and support during the long years that it took to make this book. Reilly has sunnily endured numerous "vacations" to the Silver Valley. Francesca, Grant, and Willa have explored the biking trails of the Coeur d'Alenes with their dad. All have probably heard far more than they ever wanted to about blood lead levels, lead smelters, and the Bunker Hill Company. Thank you for being such a great family.

LIVING WITH LEAD

INTRODUCTION

October 5, 1974: Idaho governor Cecil Andrus had an announcement to make. The first-term Democrat, a conservationist who had campaigned on a pledge to block an open-pit molybdenum mine in central Idaho's White Cloud Mountains, informed Idahoans that the Bunker Hill Company had reached an agreement with the state to drop its plans to sponsor a lead-zinc trade group's study of the Kellogg area's lead health scare. The company would instead support a study of the situation by the Idaho Department of Health and Welfare.[1] The governor also noted that Bunker Hill had agreed to provide "appropriate medical treatment to those affected" by the lead outbreak and that Bunker Hill would take steps to reduce its lead and particulate emissions.[2] The announcement came a month after the revelation that Kellogg-area children, in the wake of the subsequently infamous "baghouse fire" at Bunker Hill's lead smelter, had been found to have among the highest blood lead levels ever recorded. Eighty percent of 370 tested children living within 2.5 miles of the smelter tested at levels considered to be cause for medical concern and 41 of these had clinical lead poisoning.[3] Despite the seriousness and magnitude of the situation, Andrus's statement must have come as something of a surprise to politically astute Idahoans. The governor's announcement was evidence of occurrences rarely if ever witnessed in Idaho politics: publicly challenging Bunker Hill, the north Idaho mining and smelting giant that was the state's second-largest employer, and winning the confrontation.

In the weeks since news of the lead epidemic had broken, Bunker Hill had been unusually reticent, issuing only a press release expressing concern over the situation and asserting that retesting would show the State of Idaho Centers for Disease Control (CDC) blood lead results to be in error. Quietly, however, the company had been active, recruiting the International Lead-Zinc Research Organization (ILZRO), an industry-financed scientific research organization, to perform a study of the lead health situation in Kellogg. Based upon the organization's track record, Bunker Hill would have had good reason to believe that an ILZRO study likely would minimize the health implications of the lead outbreak and cast doubt upon the company's culpability in the affair.[4] Dr. James Bax, head of the Idaho Department of Health and Welfare, caught wind of the proposed study, however, and publicly blasted it, asserting that the public was unlikely to believe the results "of a study funded by lead and zinc industries."[5] A week later, Governor Andrus made his announcement.[6] The north Idaho giant had blinked.[7] While it did not grab many front-page headlines, the event marked a significant shift. Idaho, like a number of states in the 1970s, was becoming less deferential to large industry and more protective of the environment and human health.

The "environmental turn" of the 1960s and 1970s in the United States undoubtedly marked a watershed in the nation's history. The enactment of a host of major national environmental laws and the creation of federal agencies to protect the environment and worker health—prominent examples include the Clean Air Act, the Federal Water Pollution Control Act Amendments of 1972 (popularly known as the Clean Water Act), the Environmental Protection Agency (EPA), and the Occupational Safety and Health Administration (OSHA)—challenged the right of industry to pollute and to expose employees to unsafe working conditions. In Idaho, as in much of the United States, prodevelopment attitudes and values had until that time largely trumped all other concerns. In the Inland Empire—the region that is centered on Spokane, Washington, and includes much of northern Idaho and eastern Washington—the productivity of the mines, mills, lead smelter, and zinc refinery of the Coeur d'Alenes (the locale that is the focus of this work, a roughly thirty-by-ten-mile area that runs from the town of Mullan, a few miles west of Lookout Pass on the Idaho-Montana border, to the Cataldo Mission about twenty-five miles east of the City of Coeur d'Alene) were sources of pride and the subject of numerous feature stories celebrating the region's progress. If the gray, barren South Fork of the Coeur d'Alene River that flowed from the mining district or the foul, acrid smoke that issued from the Bunker Hill smelter were mentioned, it was as the necessary price for progress, a classic tradeoff for the benefits of modernization that included stable employment at decent wages for many locals. While Kellogg residents likely would have acknowledged the basic realities of their community's natural environment—the bad-smelling, acidic air; the brown lawns and barren hillsides; the "dead river" that supported no aquatic life—until the "environmental turn" it appears that most would have accepted the tradeoff of environment for economy.

Around 1970, however, attitudes about pollution and the environment began to shift, both nationally and in the Coeur d'Alene mining district. At air quality hearings in Kellogg in 1971, although many still spouted the traditional "price for progress" argument, others insisted that Bunker Hill needed to do a better job of controlling its air emissions, with some claiming that the town's bad air was the reason it had lost population over the past decade. The Kellogg High School student president, Clint Waltham, speaking on behalf of the school's six hundred students, referred to the town's air quality as a "dangerous situation" and urged "that economy give way to ecology and that pollution should be stopped."[8] The student body president was suggesting a new and radically different approach. Waltham's views, which would have been deemed odd or eccentric in the Coeur d'Alenes only a few years earlier, by the early 1970s represented the mainstream values of the Kellogg High student body, and as such, had to be taken seriously by the community.[9] As it turned out, Waltham's statement was a harbinger. For the rest of the 1970s, the Coeur d'Alenes' mining and smelting concerns, long accustomed to primacy in the region, found themselves increasingly on the defensive as they battled some environmental regulations and spent large sums to comply with others.

The transition experienced in the Coeur d'Alenes conforms closely to sociologist Ulrich Beck's theorization of two discrete stages of modernity: "classical industrial society" and "advanced modernity." Under the former, "the 'logic' of wealth production dominates the 'logic' of risk production," while "in the risk society [i.e., advanced modernity] the logic is reversed."[10] Whereas and in classical industrial society "everyone is engaged in the struggle for *his* job (income, family, little house, automobile, hobbies, vacation wishes, etc. If those are lost, then you are in a tight spot in any case—pollution or no)," in advanced modernity, "modernization *within* the paths of industrial society is being replaced by a modernization *of the principles* of industrial society . . . the gain in power from techno-economic 'progress' is being overshadowed by the production of risks."[11] The concerns with, calculation of, and efforts to minimize risk are hallmarks of the advanced modernity that has been achieved in the United States and Western Europe over the past few decades. In the Coeur d'Alenes, partially due to shifts in local attitudes and to new federal legislation and the consequent heightened scrutiny of federal and state regulators, the 1970s marked the watershed decade during which pollution and environmental risks came to outweigh traditional concerns for local prosperity and technological development.[12]

During this period, both in the Coeur d'Alenes and nationally, the truly massive levels of personal risk and environmental harm that traditionally had been offloaded by corporations onto workers, the community, and the environment came to be seen by the public as an excessive price to pay for their industrial standard of living. Simply put, people and their elected representatives began to demand a reallocation of those risks and harms so that the quality of life in places like Kellogg would be improved. This historic shift, one manifestation of which was the "environmental turn" of the 1970s, led not only to the passage of landmark environmental legislation but ultimately also to the outsourcing of much of the heavy, often dirty, industry of advanced economies such as the United States, to the lower-cost, less-regulated economies of the developing world (e.g., China), where the logic of classic industrial society still prevailed. Thus, the 1981–1982 Bunker Hill shutdown should not be viewed as an example of environment versus economy any more than the public and government responses to Kellogg's 1973–1974 lead health epidemic should be. Both incidents were part and parcel of the dynamics of industrial modernity—in this case, the shift in the United States from a classic industrial to an advanced modern society. Regardless of political persuasion, since at least the 1970s US citizens by and large have been unwilling to accept unsafe air or drinking water as the "price of progress" (as the uproar in 2016 over lead-tainted water in Flint, Michigan, evidences).

The themes of risk and tradeoffs, so central to the modern predicament, are crucial to this book. Issues of place, power, values, economics, and the environment lie at its core. This is essentially the history of a place—principally the City of Kellogg and the Coeur d'Alenes region of northern Idaho.[13] A pressing interest in questions of environmental change in the Coeur d'Alenes, involving both human and nonhuman nature, and in the complex causes of that change, is a key driver of this work.[14] Understanding the profound changes in the land that have occurred in this place over the past 130 years, of course, necessitates some study of the social and cultural forces that have been at work. As Matthew Klingle has written in his environmental history of Seattle, "place making is neither disinterested nor benign . . . place is never incidental."[15] The landscape of the Coeur d'Alenes—the product of over 130 years of mining, sixty-five years of smelting, and a thirty-plus years of environmental remediation as a federal Superfund site—is certainly neither benign nor accidental. In this place a history of economic decisions made by mining companies, a history of technological decisions regarding which processes and machinery to use in smelting, milling and mining ores, and a history of, in the words of Anne Whiston Spirn, "which groups had the power to impose certain ideas and values before weighing the consequences,"[16] are inscribed upon the landscape, and on the bodies of many who lived and worked there.

Issues of political power and the environment have been deeply intertwined in the Coeur d' Alenes. In the early 1930s a group of Coeur d'Alene City residents challenged the power of upstream mining companies to continue to pollute the lower Coeur d'Alene River and Lake Coeur d'Alene. Though influential enough to prompt the Idaho legislature to sponsor a study of the conditions of the Coeur d'Alene River watershed that resulted in the installation of some modestly effective pollution-control technology, the group ultimately lacked the power to achieve their goal. By the late 1960s, however, due to a significantly enhanced federal Clean Water Act, all of the upstream mining companies had installed the type of tailings impoundment system the Coeur d'Alene residents had advocated, and the river had begun its long journey toward recovery. Another example concerns Bunker Hill lead smelter workers. From the time of the smelter's completion in 1917, working there constituted an uncomfortable, demanding, and, due to the toxic nature of lead, potentially unhealthy job. Smelter workers, particularly after they gained true union representation in 1942, demanded safer and healthier working conditions. Until the 1970s, however, when a new federal law and a federal agency, OSHA, tilted the balance of power in their favor, the workers garnered only minimal improvements. But with federal law and OSHA backing their demands, smelter workers gained significant improvements in safety and health conditions at the plant. Shifts in cultural norms and power relations over time have profoundly shaped the physical environment of the Coeur d'Alenes.

A history of place, and an environmental history, this book seeks not only to trace significant aspects of the ecology of human and nonhuman nature in the Coeur d'Alenes, but also to explain how and why the environment of this place came to be as it is. That explanation involves untangling a number of different histories: mining and milling economics and technologies; the attitudes, ideas, and tactics of those running the district's major mining operations; labor-management relations, the thinking and strategies of the district's labor unions; and the ideologies and policy inclinations of local, state, and federal officials over time, to name a few.

A particular substance, lead, is central to this work. The heavy, ductile, versatile, highly toxic mineral plays a key role in every chapter, and one chapter is devoted entirely to "the useful metal." Having served as both an economic pillar and the toxin most responsible for devastating the Coeur d'Alenes' environment, lead's significance to the place is hard to exaggerate. This raises the question of the agency of nonhuman, and even nonsensate, actors. By traditional definition, such actors cannot be said to possess agency because they lack intentionality, and thus cannot willfully affect situations. Nonhuman actors such as lead can play important roles in history, however. Once aerosolized by a smelter and dispersed across a landscape, for instance, lead ends up in house dust and yard dirt that small children readily ingest in the normal course of play. Inside the child's body, lead goes on a path of cellular destruction that can lead to severe health problems, and even death in rare cases.

Human bodies, and their susceptibility to environmental insult, is another focus of this work. During the modernist turn in Western medicine that lasted from the late nineteenth century until the 1950s and 1960s, bodies were seen as largely impervious to their environments. Concerned primarily with the pathogens acting inside bodies, the medical profession became uninterested in the ability of environments to affect health. Post–World War II worries about radioactive fallout, pesticides, and smog, however, prompted the medical community to reconsider the environmental sources of disease.[17] The bodies of residents of the Coeur d'Alenes, and those of Bunker Hill smelter and zinc refinery workers, are central subjects of my work. Children's bodies, owing to their heightened susceptibility to lead poisoning, receive particular attention. Indeed, it was the bodies of children who lived close to the Bunker Hill smelter, as viewed through the lens of blood lead readings, that in September of 1974 alerted the world to the fact that the Kellogg area had both a severe environmental lead problem and a severe lead health issue on its hands. Until that time, environmental regulators and the Bunker Hill Company had been focused almost exclusively on the community's sulfur dioxide problem. Kellogg's 1973–1974 lead health epidemic offers a textbook example of the inextricable linkage between human bodies and local environments.

The concept of risk, and the even larger notion of tradeoffs—both inherent to the modern predicament—are fundamental to this work.[18] The systems of capitalism and industrialization that undergird modernity predicate heightened levels of social and individual risk—from pesticide-laden food to nuclear radiation—and environmental damage. Such downsides, however, are not the result of modernization's failures but of its successes."[19] Heightened risks and environmental harms are by-products of the very successes of industrial capitalism at increasing material abundance and technological sophistication that have been celebrated in such catchphrases as "the American way of life." These tradeoffs are ubiquitous. Health risks to workers in "the dangerous trades" and to nearby residents from polluting industries have been among the harsher consequences of the modern predicament.

As Joseph Stalin was fond of saying, you can't make an omelet without breaking eggs.[20] In discussing what he calls "the tragedy of development," philosopher Marshall Berman describes the inherent duality of the modern project, the impossibility of realizing the new, modern world "with clean hands."[21] While lauding "the luminous . . . possibilities that pervade modern life," and seconding Karl Marx's view that in the dynamics of capitalist development lie "a new image of the good life . . . a process of continual, restless, open-ended, unbounded growth," Berman also acknowledges that modernism's "insatiable development has left spectacular devastation in its wake."[22] One need only wander the devastated landscape of the Bunker Hill Company's old works, at the Bunker Hill Superfund site, to experience the truth of Berman's observation.

This book also deals with issues of community, focusing particularly on the small mining and smelting city of Kellogg. Because it was so beholden to its longtime chief employer and benefactor, the Bunker Hill Company (Kellogg residents long referred to the company as "Uncle Bunker"), an exploration of community in Kellogg necessarily involves an investigation of the relationship between the town and the company. With the abrupt closure of Bunker's operations in the early 1980s, however, Kellogg was forced to find a new economic rationale and identity. The community's fraught transition from a quintessentially "old West" extractive industry town to a "new West" recreation community is the focus of this work's final chapter.

Chapter 1, "Uncle Bunker's Town," sets the stage, exploring the place, the community, and the corporation where much of the action in this book takes place. The chapter traces the economic and social structure of Kellogg and the Coeur d'Alenes during the long period (1885–1982) when mining and smelting constituted the area's raison d'être. Bunker Hill's paternalistic relationship with Kellogg is examined and contextualized and the extent to which Kellogg can be seen as a "company town" is analyzed. Chapter 2, "The Useful Metal," addresses humankind's long and problematic history with lead, tracing the metal's deep imbrication in modern industrial civilization and the social costs the toxic substance has exacted. The next three chapters trace the use of various aspects of the environment as "sinks" for the mining district and Bunker Hill's unwanted wastes—their pollution. As was the rule for industry in the nineteenth and much of the twentieth century, the "ultimate sinks"—rivers, lakes, seas, airsheds—were employed first for the elimination of wastes, before the pollution began working its way back up through the food chain into human bodies. Chapters 3 through 5 describe a history that closely follows this pattern. Chapter 3, "Lead Creek," details the ecological effects of the intense mining and milling activity that has taken place along the upper South Fork of the Coeur d'Alene River and its tributaries since the early 1880s on the entire Coeur d'Alene watershed, including the Coeur d'Alene River and Lake Coeur d'Alene. The chapter describes the effects of different types of milling technology on the riverine environment, and highlights the periodic bursts of contention from downstream farmers and Coeur d'Alene City residents over the ecological problems caused by mine tailings flushed their way from the mining district. The chapter concludes with a discussion and assessment of the efforts made since 1968 to restore the ecological health of the watershed.

Chapter 4, "Foul Humours," describes the construction of the Bunker Hill lead smelter and zinc refinery in the first decades of the twentieth century and the company's perennial concerns with complaints over its sulfur-dioxide-caused "smoke damage" to trees and other vegetation. Bunker's efforts to neutralize such threats through deals with the US Forest Service (USFS) and the purchase of smoke easements from area landowners are also traced. The chapter concludes with a discussion of the company's novel revegetation efforts in the 1970s and its struggles to meet the heightened burdens of federal and state sulfur dioxide regulations—Bunker's search to make the local airshed "legible" and its ultimate decision to build the "tall stacks" as an expensive technological fix. Chapter 5, "Raining Down Poison," focuses primarily upon issues of lead health in Smelterville and Kellogg, the population centers closest to the smelter. Opening with a description of what must rank as the most dramatic episode in the environmental history of the Coeur d'Alenes, the 1973–1974 lead health epidemic, the chapter goes on to examine Bunker Hill's role in the situation. The issue of how human, and particularly children's, health has been affected by significant emissions of aerosolized lead, how this threat has been understood, and how it has been dealt with, are central to this chapter. It concludes with a discussion of the work done in the Superfund era (1983–present) to deal with the threat to children's health posed by the environmental lead that remained in the Kellogg area following the closure of the smelter.

Chapter 6, "On Lead's Front Lines," addresses the lead health issues that faced workers in Bunker's lead and zinc plants both during and, in many cases, long after their employment. In the early days of the smelter, company managers understood that lead poisoning was a danger for their workers. The chapter details the company's early approach to the issue and tracks its varied, but largely ineffectual, attempts to ameliorate workers' lead problems. It traces the trajectory of union complaint over the lead health situation at the smelter, and outlines the increase in worker protections at the lead and zinc plants following the creation of the Occupational Safety and Health Administration in 1971. The chapter concludes with a discussion of recent medical studies that have indicated health damage to Bunker Hill smelter workers, and shortened life spans, from their heavily "leaded" work environment.

Chapter 7, "A Brave New World," examines Kellogg's history as a postindustrial community in the wake of Bunker Hill's closure in 1982. It describes the community's struggle to shift from a mining and smelting town dominated by a single corporate entity to a self-reliant, Bavarian-themed ski village. Examining not only the difficulties and identity conflicts inherent in this transition, the chapter also addresses the very different economic realities that have confronted Kellogg in its new guise. It additionally traces the community's highly contentious relationship with the Environmental Protection Agency, the federal agency that since Bunker's demise has emerged as a major power and economic driver in the Coeur d'Alenes.

All this said, the question of why anyone should care about the history of a small, polluted mining and smelting region in northern Idaho—of why it matters—seems a valid one. The answer is that there is a good deal to be learned from a well-executed study of such a place. In many ways, the story of the Coeur d'Alenes is a microcosm of the modern history of the United States, and a multitude of its communities. From the time of the discovery of gold on Pritchard Creek in 1883, the forces of modern industrial society have operated with a particular freneticism and destructive capacity in this place. Home to one of the most productive mining districts in world history and to the Bunker Hill Company, for decades one of the nation's largest integrated mining and smelting operations, for 130 years the Coeur d'Alenes region has produced lead, zinc, and silver that have played an important role in powering national and global development, yielding over twenty-six billion dollars' worth of minerals.[23] As with many industrial activities, however, production has come at a significant cost. Dead swans, leaded children, a sterile river, toxic backyard dirt and house dust, denuded hillsides—these and more have been part of the place's hard reckoning. Lead from the Coeur d'Alenes was used in television sets, automobiles, gasoline, paint, computers, X-ray shields, auto batteries, and other products purchased by consumers in a global market. Although the disseminations of lead around the world via consumer goods, especially paint and gasoline, had major negative health consequences, it also helped to bring affordable products to global consumers and contributed to the vaunted American standard of living. People enjoying their cars and televisions a few decades ago in Dubuque and Dusseldorf, however, should have understood (but most likely did not) that they did not pay the full freight of their goods. This is because the pollution costs associated with mining, milling, and smelting the lead in those products was, to borrow a term from economists, externalized. Simply put, the social costs of leaded soils and people, dead horses, and poisoned rivers in the Coeur d'Alenes was passed on to the environment and human bodies to absorb.

In its 130 years the Kellogg area has been through a veritable trifecta of American modernity: intensive industrial development and environmental despoliation, large-scale environmental cleanup, and redevelopment as a recreational community. Once a strong union town filled with homeowning hard hats and busy merchants, Kellogg's shrunken population now struggles to sell lattes and condos to out-of-state visitors. The metals-laden smelter smoke is gone and the South Fork no longer runs the color of a lead milkshake. Gone too are most of the high-paying mining and all of the decent-paying smelting jobs awaiting newly minted high school graduates.[24] In its historic boom-and-bust trajectory, Kellogg's story echoes that of many communities, both in the American West, where the boom-and-bust of natural resource industries has been a commonplace, and across the globe. It is hoped that this work will both help to illuminate those shared historical patterns and assist in what social commentator and historian Mike Davis calls "excavating the future."

CHAPTER 1

UNCLE BUNKER'S TOWN

A jackass discovered the lode. In the late summer of 1885 an itinerant, middle-aged carpenter and sometime prospector named Noah Kellogg asked some merchants in the gold rush boomtown of Murray, Idaho, to grubstake him in a prospecting effort along the South Fork of the Coeur d'Alene River and up its tributary creeks, roughly thirty-five miles distant from Murray.[1] Quite possibly as much to be rid of the noisome carpenter as in expectation of enriching themselves from his finds, the merchants agreed. Provisioned with coffee, beans, other staples, and a troublesome jackass, Kellogg set out through the dense north Idaho forests. According to legend, after several fruitless days of searching for ore-bearing veins along the South Fork and Milo Creek, Kellogg decided he'd had enough and determined to return to Murray. Tired, though, he decided first to take a nap. When he awakened, the prospector discovered to his deep chagrin that his burro was missing. Perhaps following a few choice imprecations, Kellogg set off after his missing ass. After trudging up one steep hill and down another, the weary carpenter finally came across his burro munching upon some flowers and grass on a hillside. Sitting down to enjoy his pipe after all the exertion, Kellogg looked down and saw a large outcropping of galena (silver-lead ore). Kellogg's ass had led him to one of the richest areas of lead-silver mineralization in the world.

Within a few years, near the site where Noah found his wandering jackass, the town of Kellogg, and to an even greater extent its up-canyon neighbor, Wardner, had emerged as bustling centers of mining development. In contrast to gold panning and other types of placer mining, as a rule silver and lead mining require significant investments of labor and technology. Silver and lead usually are found mixed in with other ores, and often hundreds of feet below the earth's surface. Costly quartz mining, as well as milling and smelting processes, must be applied in order to extricate the valuable minerals from the waste rock with which they are entangled. Limited liability corporations, with their generally unrivalled capacity to marshal technology, labor, and capital, have dominated the field. By the late 1880s some of the biggest names in American finance and mining engineering could be found in Kellogg and Wardner, working to expand their fortunes. The mining district's largest operator, the Bunker Hill and Sullivan Mining and Concentrating Corporation, had the backing of a list of luminaries that included Simeon Reed, a Portland, Oregon–based railroad and iron magnate; John Hays Hammond, the world-famous mining engineer; San Francisco bankers William H. Crocker and D. O. Mills; and a Chicago syndicate headed by banker James Houghteling that included iron and steel magnate Edward Ryerson and farm equipment manufacturer Cyrus McCormick. Even if all the silver and lead lying in the Coeur d'Alenes could be efficiently mined and refined, however, it still would remain worthless unless it could be transported at a reasonable cost to far-flung markets. By 1891, a mere six years after Noah Kellogg's lucky find, that problem too had been solved with the construction of two competing railroad lines into the narrow Silver Valley—a sure sign of the promise the Coeur d'Alenes' mineral deposits evinced for major Gilded Age capitalists.

Despite its picaresque beginnings, however, the town of Kellogg took some time to emerge as the leading community in the Coeur d'Alenes. Kellogg was dwarfed by its immediate neighbor, Wardner, until 1903, when the Bunker Hill Company, better known to locals as "The Bunker" or "Uncle Bunker," completed a tunnel moving the entrance to the sprawling Bunker Hill Mine two miles northwest, thus connecting the town to its underground mine workings. The company spent six years and considerable resources excavating the Kellogg Tunnel, which allowed the company to discontinue use of the unreliable and inefficient aerial tramway system it had used to transport ores from the mine to rail-loading facilities.[2] As miners and their families moved "down the hill" from Wardner to Kellogg—situated predominantly on the south side of the narrow valley carved by the South Fork of the Coeur d'Alene River—businesses, saloons, schools, and churches followed.

Although named for the prospector whose jackass happened upon the great lode of galena, Kellogg's destiny would be shaped primarily by the mining and ore processing corporation that brought the miners "down the hill" in 1903—a company that came to be known to locals simply as "Uncle Bunker."

The Bunker Hill Company, too, came to identify itself closely with the growing town of Kellogg. Bunker Hill letterhead for some years identified the company with both Wardner and Kellogg, but by 1907 the reference to Wardner had been dropped.[3] In 1914 the Union Pacific railway changed the name of its local depot from Wardner to Kellogg. The town became home to Bunker's management office, much of its means of production, and many of its workers. Kellogg offered easy ac-

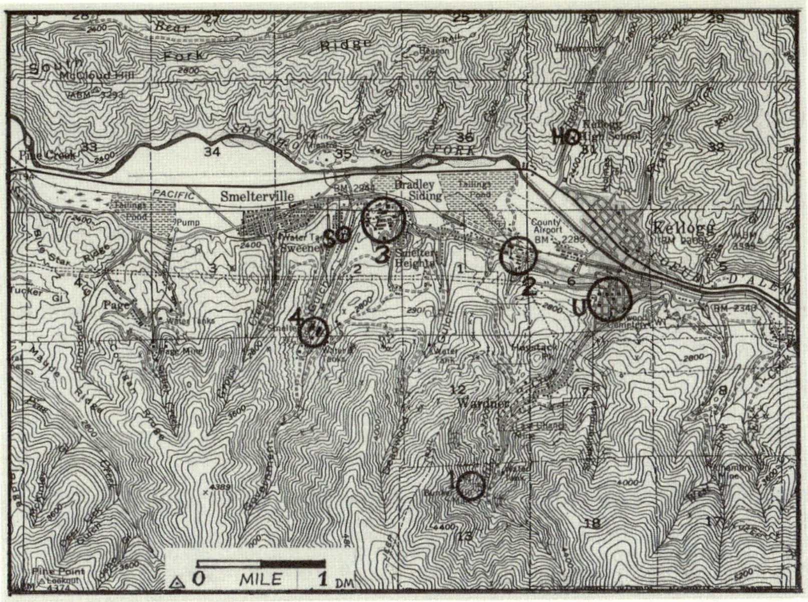

FIGURE 1.1. Kellogg-Wardner Portion of a 1957 USGS Map of the Kellogg, Idaho area (credit: Dale Martin). Area of Kellogg and Bunker Hill Company properties, located on US Geological Survey 1957 topographic map. Contours are separated by eighty feet of elevation. Key to Bunker Hill properties: 1. Bunker Hill Mine upper entrances; 2. Bunker Hill Mine tunnel entrance, mill, and mine office; 3. Smelter; 4. Zinc plant. Key to Kellogg features: U, Uptown business district; S, Silver King Elementary School; H, Kellogg High School.

cess to the railroad service that moved the company's ores to distant smelters and markets. Located in a valley, albeit a narrow one, Kellogg had room to expand to provide housing and services for additional workers. By 1928 Bunker Hill's industrial complex would include, either within or closely adjacent to Kellogg's city limits, crushing mills, settling ponds, a major lead-silver smelter, a large zinc refinery, and the massive Bunker Hill Mine. With so much at stake in the small north Idaho mining and smelting community, it is no wonder that Bunker Hill took a strong and proprietary interest in Kellogg's development.

A Company Town of a Sort

While not a company town in the strictest sense, Kellogg in many respects resembled one from 1903 until the demise of the Bunker Hill Company in 1982. In some of the purer examples of company towns, such as Pullman, Illinois, or Potlatch, Idaho, the corporation disallowed local politics and directly ran the communities, along with a large company store.[4] With the rise of industrial capitalism, a number of company towns sprang up around the world from 1830 to 1930.[5] While utopian wishes and concerns for social engineering colored a number of company towns, corporate desires to secure an ample, quiescent, nonunion workforce was in almost all cases a hallmark of these social experiments.[6]

Kellogg resembled a company town in several respects: Bunker Hill dominated the local economy and much of the town's population owed its livelihood directly to the company; until the 1950s, the company owned a considerable amount of local housing and rented and sold houses to many of its employees; Bunker Hill's property tax payments constituted a major share of local tax revenues; and the company's various forms of corporate welfare provided significant amenities to Kellogg residents. In other ways, however, Kellogg differed from a pure company town. For instance, a considerable amount of Kellogg real estate was not owned by Bunker Hill. The company was not highly involved in city planning and did not attempt to impose a uniform architectural style in the town. Bunker did not oppose Kellogg's successful early twentieth-century efforts to incorporate and does not appear to have meddled much in city politics. And, as the mining community's protean drinking culture attests, the company did not waste its time trying to turn Kellogg into a dry community. Still, while Kellogg had more autonomy than a pure company town, the predominant role played by Bunker Hill in the community's economic life ensured that it remained a species of company town.[7] Wendell Brainerd, editor of the *Kellogg Evening News* from 1945 through 1985, stated revealingly, "We're not going to come out against Bunker Hill. Bunker Hill is Kellogg. Kellogg is Bunker Hill."[8]

In turning Kellogg into a type of company town during the first decades of the twentieth century, Bunker Hill executives F. W. Bradley and Stanly Easton were following an antiunion corporate strategy that already had proven its worth in the turbulent coalfields of southern Colorado. As Thomas Andrews shows in *Killing for Coal*, his groundbreaking study of the Colorado coalfield wars, coal company operators took note of the fact that the major "marching strike" of 1894 broke upon the shoals of the company coal towns—places such as Rouse and Berwind—where strikers proved unable to enlist sufficient numbers of colliers in their cause. Whereas marching strikers garnered strong support in the "free-standing open camps" where most miners had built or owned their own homes, the marchers were far less successful in the closed camps that had been built by the coal companies and where workers generally rented housing or ground from the companies.

Mining executives in the Coeur d'Alenes, beset by strikes and outbreaks of union violence throughout the 1890s, undoubtedly were well aware of the success their Colorado colleagues had seen with the deployment of the company town. Bunker Hill's Bradley and Easton thus attempted to forge in Kellogg a procompany "family town" that would form a bulwark against what they perceived as the radical unionism that flourished in the saloon-ridden "vernacular landscapes" (in the language of J. B. Jackson) of such nearby open camps as Wardner and Burke. Various, and often highly generous, forms of corporate welfare would be fundamental to the company's strategy of making Kellogg a stable, loyal, working-class community.

The Makeup of Kellogg

In 1890, Kellogg's population was only 324. A decade later, the town counted 823 residents, a paltry number compared to the 2,278 people living in Wardner or the 2,265 residing eleven miles east in Wallace, the Shoshone County seat.[9] By 1910, however, following completion of the Kellogg Tunnel, Kellogg's population had nearly doubled to 1,552, while Wardner's had declined to 1,369.[10] In 1920 Kellogg's population surpassed that of its intracounty rival, Wallace, and never looked back. Kellogg's population reached its zenith in 1957, when it climbed to a size of approximately 5,900.[11] Then, during the seven-month strike of 1960, the small city lost nearly a thousand residents.[12] Kellogg continued to decline, dropping to a population of 3,417 in 1980, shortly before the bottom fell out of the local economy in 1982 with the closure of Bunker Hill's operations. A population of 2,227 was recorded for 2000, the lowest population total since the 1920s in the onetime mining and smelting town.[13]

Throughout its history, Kellogg has largely mirrored the broader demographic patterns of Shoshone County. Of Kellogg's 4,636 residents in 1930, 2,620 (roughly 57 percent) were male and 894 (approximately 19 percent) were foreign born.[14] An additional 1,224 (26 percent) were born of one or more foreign parents.[15] There have been relatively few black, Asian, or Hispanic residents of Shoshone County during its existence.[16] The 1890 census reported the presence of 44 "colored" people in a total county population of 5,179, but by 1930 that number had plummeted to 6.[17] Still, Kellogg, like Butte, Montana, and other western mining towns, traditionally has boasted a rather wide cross-section of European nationalities. This is supported both by the census data and the recollections of Julie Whitesel Weston, who grew up in Kellogg in the 1940s and 1950s. Weston's memoir notes the presence of émigrés from Finland, Wales, Poland, Croatia, England, Mexico, and, following the failed 1956 uprising, Hungary. Most vividly, Weston experienced her hometown as "full of big Italian families."[18]

Labor Strife Molds the Bunker Hill–Kellogg Relationship

Bunker Hill Company President F. W. Bradley was scarred, both literally and figuratively, by the labor wars of the 1890s. Bunker Hill originally brought Bradley, who graduated from the University of California, Berkeley in 1885 with a degree in mining engineering, to Kellogg in 1890 to serve as assistant mine manager. In 1893, Bradley became mine manager, and in 1897 he ascended to the company presidency, a position he would hold until his death in 1933. On November 17, 1904, a bomb ripped through the front door of Bradley's home in San Francisco.[19] The blast destroyed the entire front of the house and threw the mining company head out onto the street. Deaf and blind for two months following the explosion, Bradley carried scars on his face and body for the rest of his life.[20] Western Federation of Miners (WFM) member Harry Orchard later confessed to planting the bomb that nearly killed Bradley. Orchard claimed that the WFM carried out both the attempt on Bradley's life and the successful assassination of former Idaho governor Frank Steunenberg a year later in revenge for the union's defeat during the 1899 labor war in the Coeur d'Alenes mining district.[21]

Like many parts of the country during the last quarter of the nineteenth century, a period of frenetic industrialization in the United States, the Coeur d'Alenes were wracked by intense strife between labor and management. The area's first labor violence took place in 1892 in Burke Canyon, a heavily mined area lying roughly fifteen miles northeast of Kellogg. The announcement of districtwide wage cuts, effective April 1, 1892, prompted miners to go on strike. Mine owners responded by hiring scabs. On July 11 company guards and union men at the Gem Mine and the Frisco Mill exchanged shots, and a dynamite explosion destroyed part of the Frisco Mill. When the smoke had cleared five men were dead—three union members, a detective, and a nonunion worker. The following day Idaho governor Norman B. Willey declared martial law in Shoshone County. The military then constructed "bullpens" to incarcerate all those deemed a threat to public order.[22] Discovering himself surrounded by a large union force during the disorder, Bunker Hill manager Victor Clement promptly offered to close down Bunker Hill Mine.[23]

Tensions continued in the district throughout the 1890s. On July 10, 1895, fire destroyed the houses of three local businessmen known to support Bunker Hill, considered by unionists to be "the last bastion of managerial hegemony in the region."[24] An explosion at the company's flume, coupled with a fire at the mill, took place on May 10, 1896. Though the apparent effort to destroy the Bunker Hill mill proved unsuccessful, destruction of the flume might well have prevented the company from effectively fighting the mill fire.[25] In 1899, labor violence in the Coeur d'Alenes reached a peak. As the only mine operator in the district paying below the standard rate of $3 per hour to shovelers and $3.50 an hour to miners, Bunker Hill was a natural target for union outrage. On April 29, 1899, organizers lost control of a labor protest. Union men hijacked a Union Pacific train at Burke and drove it to Wardner, where they exchanged gunfire with a Bunker Hill watchman. Bunker's ore concentrator, valued at a quarter of a million dollars, was dynamited and completely destroyed; the company's Kellogg offices were burned to the ground. The violence left one union and one nonunion worker dead.[26] Steunenberg, a Democrat and an ally of the mine operators, immediately declared martial law in the district. On May 4, federal soldiers took control in the Coeur d'Alenes. They began building bullpens and liberally incarcerating not only unionists but also local People's Party officials and members, who were believed to be union supporters.[27]

Although Bunker Hill quickly rebuilt its mill and fully resumed its operations, F. W. Bradley and other top company officials remained deeply affected by the labor strife of the 1890s. For decades, a principal lens through which Bradley and Stanly Easton, Bradley's longtime right-hand man, judged workers and local citizens was by where they stood in 1899—with or against the company.[28] Perennially antiunion, Bradley believed the WFM to be the worst of a bad lot, often referring to the organization simply as "dynamiters." The Bunker Hill chief viewed the prounion Populists and Populist-Democrats, who had gained control of numerous Shoshone County offices during the 1890s, as nearly as harmful and dangerous as their union allies. Bradley, a lifelong Republican, followed Shoshone County politics closely and almost invariably directed company campaign contributions to what he perceived as safe, law-and-order GOP candidates.

Following the troubles of 1899, Bunker Hill adopted a two-pronged approach to labor and community relations. On the labor front, the company adopted a hard, antiunion stance. Joining with other district mining companies and working through the Mine Operators Association (MOA), Bunker helped to establish and finance an employment office responsible for screening all potential employees to ensure that no WFM members were hired in the district. In order to receive a permit for employment in the Coeur d'Alenes, a worker had to take an "ironclad oath" foreswearing union membership. Company physician and county coroner Dr. Hugh France served as the ultimate gatekeeper in this process. It was theoretically impossible for a worker to gain employment from a mining company in the district without a permit signed by France.[29] A common source of complaint in letters between Bradley and Easton concerned the failure of other mining companies in the district, and particularly the Guggenheim-controlled Federal Mine, to faithfully use the employment office for all their hiring. The Bunker Hill chieftains looked askance at examples of such negligence, worried by the possibility that it would permit the scourge of labor radicalism to reappear in the mining district.[30] Another aspect of Bunker's hard approach to labor relations was its hiring of Pinkerton detectives, from the 1890s through at least the first decade of the twentieth century, to infiltrate local labor organizations and report on their plans and activities.[31]

The other prong of Bunker Hill's approach, however, involved development of corporate welfare efforts. This stratagem derived from essentially the same motive as did Bunker's blacklisting efforts—to keep radical labor groups such as the WFM and the Industrial Workers of the World (IWW) from gaining a foothold in the Coeur d'Alenes generally, and in the company's operations specifically. Bunker's corporate paternalism represented the company hierarchy's fundamental desire to foster a loyal and quiescent workforce by directly associating the company both with activities that directly benefited employees and with actions that made Kellogg a more desirable place to live.

Andrea Tone, a scholar of American corporate welfare, or "welfare capitalism," as she refers to it, identifies the strategy as a fundamentally conservative business response to the public welfare proposals of the Progressive Era (1890–1920). Supporters of welfare capitalism contended that the opportunities it provided to (male) employees to "earn" pensions, stockholding plans, and other forms of property were inherently preferable to the intrinsically emasculating handouts provided by state welfare programs.[32] The system caught on quickly. By 1914 over 1,500 US firms were engaged in some form of welfare. The lion's share of these corporations was large—79 percent of them had over a thousand employees and the average firm practicing corporate welfare employed more than 7,500 workers.[33]

Corporate welfare took a wide variety of forms, including the ten-hour day for women; thrice-daily "rejuvenating" calisthenics breaks; "free" hot lunches; lavishly decorated rest areas; health insurance, sick benefits and pensions; roof gardens; health and occupational safety plans, and stock-sharing plans.[34] Though undoubtedly there was genuine goodwill involved in some corporate welfare efforts, most practitioners seem to have viewed welfare primarily as a sound business decision. Among corporate welfare's broad goals were the recruitment and retention of a labor force, securing heightened levels of labor productivity, and diminishing the likelihood of labor unrest.[35] Employers believed that corporate welfare efforts would create a more satisfied, quiescent workforce, in part by addressing the problem of "corporate estrangement" from workers that had developed with the rise of large, impersonal corporations.

Bunker's long-standing practice of a wide range of corporate welfare strategies largely coincides with national trends in this arena. Bunker's politically conservative management team of Bradley and Easton no doubt welcomed corporate welfare's anti-Progressive, antistatist bent. Coming directly on the heels of the Coeur d'Alenes' labor wars of the 1890s, however, the principal motives behind Bunker's welfare efforts were the maintainance of a nonunionized workforce and labor peace. In addition, Bunker's corporate welfare programs far outlasted those of the vast majority of its corporate contemporaries. While most corporate welfare programs dropped all but their pecuniary benefits by the late 1920s—particularly insurance, stockholding plans, and pensions—Bunker Hill continued many of its forms of welfare until the early 1980s.[36]

The "Company Union"

Part of Bunker Hill's corporate welfare policy involved lavishing support on the Wardner Industrial Union (WIU), the "company union" that Bunker established in the wake of the 1899 debacle. For Bradley, the company-sponsored industrial union acted both as a kind of insurance policy and as a litmus test, simultaneously protecting the company against labor violence and identifying which workers could be trusted. The Bunker president wanted to enroll all company workers as members of the industrial union, since "we can take it for granted, that all employees who are not members of it, are not heartily in sympathy with us; that is, that if there are a hundred Western Federation men in our mine, they are surely not members of the Wardner Industrial Union."[37] At the WIU's inception, in September 1899, Bunker Hill manager Frederick Burbidge proposed that the company would match employee donations to the union on a monthly basis.[38] When in 1905 Bunker sent the WIU a check for $1,257.25, Easton offered his hopes that the union would "continue to prosper and I would be very glad to hear of an increase in its membership."[39] Although during its decades of existence, Bunker Hill would claim that the WIU was a legitimate bargaining agent, there is little evidence that the ersatz union ever did more than submit mostly toothless requests to the company.[40]

In another instance of Bunker Hill's paternalism, in an era long before Social Security and Medicare, Bradley wanted to ensure that "employe[e]s who have grown old in our service and who have made good" did not go destitute. He therefore proposed that such workers be given continued employment by the company as fire watchmen or as janitors, or that they could continue mining in "some easy ground" to lease in the Bunker Hill Mine, "whereby they may safely make a living."[41] Bradley's concern for the welfare of loyal older workers was part of his ongoing determination to reward "good employees" and to cultivate a cooperative workforce.

Corporate Beneficence

Largely from a desire to secure good community and workforce relations, Bunker engaged in a host of efforts, most centered on Kellogg, aimed at improving the quality of life in the blue-collar community. The company provided electric power and water to the communities of Wardner and Kellogg from the early 1900s until well after World War II, built the first schoolhouse in the area and deeded it to the community, and leased, and later donated, land on its property for Kellogg's second school, the Lincoln School. Bunker also constructed the first and second hospitals in the area, the first in Wardner and the second in Kellogg, and donated the land and water system for the third hospital, the Shoshone Medical Center, built in 1957–1958. For decades the company also ran boardinghouses for single employees, in both Wardner and Kellogg.[42] Bunker Hill made donations to churches, thinking them a good influence on local culture; hired scores of local youths for summer work; and from 1952 to 1981, in tandem with the Hecla mining company, gave out hundreds of college scholarships that helped local students attend Idaho colleges and universities.[43] For a number of years, the company also owned the local newspaper, the *Wardner News*.[44] When in April 1917 Bunker was faced with a US Postal Service requirement that as the newspaper's owner it publish a list of all stockholders with over 1 percent of its stock, the company sold the paper to its publisher, W. L. Penny. Easton informed Bradley, however, that there was no cause for concern over this development, because "our relations with the paper will be just the same, having a mortgage on it and being its chief patron in the way of job printing, as though the outfit was owned by us as at present."[45]

The company's greatest contribution to the community, however, was its construction and maintenance of a YMCA-administered "industrial recreation facility" in uptown Kellogg for company workers and community members. The Y would be a community fixture for nearly seventy years.[46] On February 1, 1910, Bradley authorized Easton to proceed with construction of the two-story YMCA building "on the basis of cost of $26,000, although I presume this means, with furnishings, an ultimate cost of what I hope will not exceed $30,000."[47] The redbrick facility opened to the public in December 1910. A formal dedication ceremony in February 1911 included members of the company's board of directors and prominent stockholders. Bunker leased the top floor of the building to the Wardner Industrial Union.[48] Eventually offering club rooms, reading and library facilities, a gymnasium, swimming pool, showers, bowling alleys, and other recreational equipment, the facility bore some resemblance to an urban gentleman's club or a community college.[49] Over the years, Bradley and Easton corresponded extensively about the YMCA and seem to have taken substantial pride in the facility.

Bunker Hill treated the administration and maintenance of the YMCA facility as a normal cost of doing business, despite the fact that the venture always remained a money loser.[50] In addition to constructing the YMCA building, Bunker contributed substantially to the annual maintenance of the facility and paid the YMCA director's salary.[51] In October 1917, Easton suggested to Bradley the advisability of adding a third story to the YMCA building, "for lodge and other social meetings," at an estimated cost of up to $10,000.[52] The new floor's rooms were to be leased and furnished by the WIU. In approving Easton's idea, Bradley mentioned his belief that construction of the third story would "add another favorable 'tie' in our relations with our own crew and the community."[53] In a similar vein, in 1918 Bunker began construction of a Staff House, located directly across McKinley Avenue from its corporate offices. The Staff House contained temporary lodgings for miners and managers recently hired by the company and also provided space for social events such as wedding receptions, tea gatherings, and Christmas parties for local businesses.[54]

In October 1924, the Kellogg YMCA, with company backing, opened a night school. Boasting a faculty of at least six, and offering courses in citizenship, English, geology and mineralogy, mathematics, and mechanical drawing, the school charged only nominal fees and was open to all adult community members. Easton was optimistic that enrollment would reach one hundred for the school's first term.[55] By the late 1930s the school's enrollment was averaging between two hundred and three hundred per year, with roughly equal numbers of men and women taking classes. Credit for the school's heightened popularity probably lay with its increased emphasis on vocational education—by the late 1930s courses in determinative mineralogy, carpentry, machine work, drafting, automobile mechanics, and shop mechanics were included among its vocational offerings. The school continued to offer some "purely academic and cultural" courses, however, and, in order to "Americanize" the relatively large number of foreign-born workers hired by Bunker, citizenship and remedial English classes.[56]

Bunker Hill's "industrial YMCA" appears to have filled a niche in Kellogg and outlying Coeur d'Alenes communities. In 1938, a registered YMCA membership of 1,200 and a user community of 2,000 enrolled in nearly 200 organized activity and program clubs. Many of these were youth groups, led by roughly fifty volunteer group leaders. Among the youth groups was Hi-Y, the national organization's high school fraternity, which in Kellogg catered to both girls and boys. In the brief north

Idaho summer, residents of the Coeur d'Alenes turned their attention to outdoor recreational opportunities. The YMCA accommodated their interests, sponsoring baseball, softball, and other sports leagues; an aquatic program at the city's outdoor pool; and a summer camping program for boys and girls at a mountain campsite twenty miles from Kellogg.[57] In 1916 Bunker Hill donated the land and money to lay out the YMCA's baseball fields, bleachers, and playground equipment, an attractive recreational assemblage that anchors uptown Kellogg to this day.

Kellogg and the "Industrial YMCA" Movement

Kellogg was one of many American communities to boast an industrial YMCA. In the 1870s the YMCA and railroad executives came together in an effort to combat the problem of an increasingly restive, class-conscious workforce. The result was the creation of "railroad YMCAs." These facilities, located along the railroad lines, were meant to offer, like the company town and other forms of corporate welfare, a moral, uplifting alternative to saloons and brothels. In the words of Thomas Winter, the historian of the YMCA's work in this field, "railroad YMCAs . . . provided a place for leisure and for time spent between 'runs,' offering overnight accommodations and meals—all in combination with non-denominational religious work."[58] In 1903, at the behest of manufacturing concerns, the YMCA expanded its programs to urban, industrial workmen. By 1906, the organization's recently established Industrial Department had established industrial associations in twenty-four locations, spanning eight states. By 1921, at the movement's high-water mark, 218 YMCA officials served workers in urban industries and 226 catered to the needs of the nation's railway workers. Companies often paid for an association building. Bunker Hill's costs in this regard were in the middle range; in 1905, a typical YMCA building usually cost between $10,000 and $50,000.[59] In addition, corporations and local membership dues needed to pay for the facility's upkeep and the salary of a local YMCA secretary, a trained YMCA employee who ran the facility. For their money, corporations could expect the YMCA to provide "an ideal of Christian manhood to the workers. . . . Once the workers adopted a higher ideal of manhood, rooted in values of Christian brotherhood and service . . . workingmen would abstain from political radicalism and labor unrest."[60] Despite their appeal to corporate managers, the industrial YMCAs also held an attraction for many workers, who as a rule were expected to contribute to their maintenance. Winter notes that "in the name of social harmony and respectable manhood, craftworkers felt that a union would have to exercise certain edifying and uplifting influences over its members. . . . YMCA programs . . . [involved] them in a web of uplifting activities, all of which were designed to make them better men and more loyal employees."[61] As was the case in Kellogg, workers commonly participated in governing the local YMCA. Local support for the industrial YMCAs was often strong—it was in Kellogg—even if workers often interpreted the meaning of the institution differently than did company bosses.

Industrial YMCAs were underwritten by essentially the same broad goals and purposes that, at roughly the same time, inspired other forms of corporate welfare and company towns. All were part of a project aimed at undermining a masculine working culture, centered on the workplace and the saloon, that enshrined mutualism and solidarity, and placed a premium upon a masculine assertiveness toward the boss. The industrial YMCA sought to supersede that workingman's culture with a culture in which responsible, sober, Christian manhood would bond together workingmen and their corporate managers as Christian brethren and men of character. In this relationship, the familial, sentimental values of the YMCA would dominate workers' relations with their corporate superiors, while company officials would be free to follow the logic of the marketplace in their workplace relationship with workers.[62] Although YMCA officials may have wished to "harness the forces of industrialization to their own project, creating a society in which Christian manhood would serve as a matrix for social order—a society in which moral, pious men, regardless of social standing, would self-sacrificingly cooperate in the production of industrial wealth, while bringing about the kingdom of God on earth," it is doubtful that many workingmen were seduced by this vision.[63]

By the 1920s, however, the railroad and industrial YMCA movements were running out of steam. After 1920, many companies curtailed their support for YMCA's industrial welfare services and shifted their support to trained industrial relations experts, sometimes setting up in-house Departments to handle this aspect of employee relations.[64] Bucking the trend, however, Kellogg's industrial YMCA persisted as an integral community organization, with strong support from Bunker Hill, into the late 1970s.

Housing for Bunker Employees

For much of the twentieth century, a remarkable feature of life in Kellogg was the fact that Bunker Hill employees could rely upon the company to take care of their housing needs. The most fortunate were supplied, gratis, with houses owned by Bunker, others rented at low rates, and sizable numbers utilized low-interest company financing to purchase homes from the company. In a feature on Bunker published in 1921, the eminent mining writer T. A. Rickard described the company's policy to encourage homeownership among its employees, explaining that Bunker leased up to an acre of land to employees for a dollar a year, lent them the money to build or purchase a home at 6 percent interest when the banks charged 8 to 12 percent, or lent them the materials to build their own homes at the same 6 percent rate of interest.[65] Recalling his experiences as a young miner in the 1920s, Julian Marshall gave personal credit to Stanly Easton for enabling him to secure a home in Kellogg at little expense, noting, "In 1922, [Easton] gave me a lease on a fifty-foot lot at W. 624 McKinley Avenue in Kellogg on which I built a five-room frame house for $3,100. I didn't have to put a dime down. The Bunker Hill financed the entire deal and my monthly payments were nominal."[66]

Bunker Hill remained heavily involved in the Kellogg housing market until the mid-1950s, when, under the direction of new company president John D. "Jack" Bradley, the company put much of its housing stock up for sale.[67] In June 1956, the company announced that it was going to sell 80 single-family residential units and begin renting its remaining 160-odd properties for close to market rates.[68] The houses put up for sale were offered first to their residents, who, if they did not wish to purchase, were permitted to rent from the company at increased rates. Bradley portrayed Bunker's move out of the housing business as a means both to promote "free enterprise real estate development in Kellogg," which he believed had long been stifled by the company, and to enable Bunker to focus greater attention on its primary business concerns.[69]

At the same time that Bunker Hill was retreating from its longtime housing policy, the company also placed its electric utility, the Lighthouse, and a related appliance store business on the auction block. The company-owned utility had served the communities of Kellogg and Wardner since the early years of the century, but to a new generation of Bunker leadership represented by Jack Bradley it was a relic of a bygone era and an unwanted corporate appendage. In June 1956 Bunker sold the Lighthouse to the Washington Water Power Company for about $805,000.[70] Announcing the company's decisions to divest itself from the housing and utility businesses in Kellogg, general manager Wallace Woolf suggested a fundamental shift in Bunker's relationship to the town. While insisting that the company "is in business to stay, and in Kellogg to stay, as well," Woolf added that Bunker Hill "should be in the position of just another 'citizen'—a very interested one, to be sure—and not a paternalistic organization which allows no room for the city to cultivate its own self-respect and initiative."[71] The prosperous mid-1950s represented a period of rapid change and development for Kellogg. The small city was in the midst of a preparing a master plan for long-range development, along with projects to build a new hospital, high school, and airport. Bunker Hill strongly supported the city in all these endeavors, albeit in its new, less paternalistic role.

The Company and Local Governments

Taxes had long been a critical part of the relationship between the city and the company. Shortly after Kellogg incorporated in 1907, Easton reported to F. W. Bradley that "I [e]nclose [for] you [a] small map showing the portion of our property included in the Kellogg incorporation. I have excluded therefrom all portions of our property on which a heavy assessment can be levied, but we have enough people and enough holdings within the town limits to give us a good voice in the conduction of its affairs and government."[72] Notwithstanding his efforts to keep Bunker's local tax burdens "very light," Easton expressed the need to maintain "friendly relations" with the town and ensure "that its officials be not antagonistic to the company."[73]

In 1913, the Kellogg City Council passed a resolution requesting that Bunker Hill permit the extension of city boundaries so as to allow the incorporation of the company's concentrator and other facilities into the city. The obvious purpose of this move was simply to increase the city's tax base. The very fact of the request, however, evidenced the city's comparatively weak position relative to the company. In discussing the proposed incorporation with Bradley, Easton noted that it would entail a significant hike in Bunker's taxes, from the $181.44 the company paid the previous year to $4,559.17.[74] Despite Bradley's misgivings about the added tax burden, the company acquiesced to the incorporation of the concentrator and other equipment.[75] In addition to its other local tax burdens, by the early 1920s Bunker was paying 60 percent of Kellogg's school taxes.[76] In the mid-1950s the company granted Kellogg another tax windfall by not contesting the incorporation of its mine and smelter within the city limits.[77]

By 1913 Bunker officials believed that Kellogg had grown large enough to warrant a sewer system. Bunker had a sewer system for its operation but lacked a city system with which to connect. When the City of Kellogg encountered difficulty selling its sewer bonds, Easton and Bradley took the matter up with their company banker, William Crocker, a San Francisco financier who was also a major Bunker Hill stockholder. Bunker purchased the city's sewer bonds, at a cost of $23,000, and thus enabled the construction of Kellogg's sewer system.[78]

Bunker Hill and Kellogg during World War I

With the entry of the United States into World War I in April 1917, Bunker Hill spearheaded Kellogg's efforts to support the war effort, with General Manager Easton and his wife, Estelle, personally leading much of this work. During the war, Estelle Easton acted as county chair of the Red Cross, head of the War Insurance Board, and chief of War Relief Work.[79] To encourage worker purchases of Liberty

Bonds, the company offered a plan that permitted employees to buy a hundred dollars' worth of bonds for only a ten-dollar down payment. The company fronted the remaining ninety dollars and then recollected it, with no interest, over the subsequent four months via payroll deductions.[80] Easton soon proudly reported to Bradley that for the first round of Liberty Loans, 213 employees subscribed through the company for a total of $21,450, while another 76 made subscriptions totaling $6,080 through local banks. When combined with the subscriptions of Bunker management and their families, the total climbed to $41,530.[81] As a gesture of support for the war effort and goodwill toward local families who had members serving in the conflict, in July 1917 the company announced its cancellation of all rental, electricity, water, and building loan interest charges to the families or dependents of its employees who were serving in the armed forces.[82]

Worried about Wobblies

During World War I, Bradley and Easton, along with other Coeur d'Alenes mining chieftains, were much alarmed by reports of local activity by the Industrial Workers of the World (IWW)—more popularly known as "the Wobblies."[83] The radical labor organization, a successor to Bunker's old nemesis, the WFM, was seen as "brazenly pro German" by Bradley and others.[84] In the highly charged atmosphere following the American entry into the war, the IWW, with its inflammatory anarchistic rhetoric and antiwar stance, became a ripe target for corporations, the media, and the US government. In one of the more notorious incidents of the period, in response to an IWW-led strike at three Bisbee, Arizona, copper mines, on July 12, 1917, a *posse comitatus* arrested more than two thousand men, most of whom were strikers. Later that day the deputies forced 1,186 of them onto a train and unloaded them at Columbus, New Mexico, 173 miles to the east.[85] In another infamous episode, during the early morning hours of August 1, 1917, shortly after giving a fiery antiwar speech, IWW organizer Frank Little was abducted from his Butte, Montana, boardinghouse room, dragged behind a car, and hung from a railroad trestle. His killers were never discovered.[86]

The IWW indeed was active in the Inland Empire and the Coeur d'Alenes in 1917, playing upon wartime labor shortages to organize timber workers and miners in demands for higher pay and better working conditions. A large IWW-led timber strike proved threatening enough to prompt Idaho governor Moses Alexander to move his official residence to Coeur d'Alene for the summer of 1917 in an effort to break the strike.[87] In the Coeur d'Alenes, Easton believed that IWW-influenced "roughs and renegades from Butte" had filled "all the mines east of Kellogg, except the Hecla."[88] In April, in response to "some report of seditious talk," two additional watchmen were put on duty and "a lot of new electric lights installed at strategic points" at the lead smelter, then still under construction.[89] Discussing the IWW situation in July, Easton informed Bradley that "an organized financed campaign is being systematically carried on to tie up all industries hereabouts. . . . These agitators and pickets are reported to be getting $3 per day and are surely being supported by outside (German) money."[90] Noting that the IWW had made inroads in the upper part of the district, Easton reported that in Kellogg, "We are extremely careful . . . and yet have no cause for alarm."[91]

The quiet situation in Kellogg was due in no small part to the aggressive anti-IWW stance taken by the city government, undoubtedly with Bunker's full support and encouragement. In early July the city swore in one hundred extra police officers. Announcing that Kellogg was willing to swear in five hundred extra officers if necessary, Mayor T. R. Mason stated baldly that "the I.W.W. organizers will not be permitted to hold meetings here. Nor shall the I.W.W.s get a footing here."[92] Easton informed Bradley at this time that there were in Kellogg "I.W.W. agitators who are planning to start trouble here," and moreover that "their speakers on the streets of Spokane say the [newly opened] Bunker Hill Smelter is to be their special point of attack."[93] When the dust from the Wobbly scare cleared, however, it was evident that both the City of Kellogg and Bunker Hill's operations remained unscathed. Bradley credited the bonds Easton had forged between the company and the Kellogg community with dimming the IWW's local appeal, writing him, "If we were not treating our own crew and the people of Kellogg fairly and decently, these [IWW] circulars might bear some fruit. But the Liberty bond and Red Cross subscriptions on the part of our crew indicate to me that you have tied the Bunker Hill & Sullivan organization mighty well with both crew and the community."[94]

The loyalty of its workforce and the militant support of Kellogg's elected officials during the IWW agitation of 1917 offered a marked contrast to the situation Bunker Hill had faced during its struggles with the WFM in the 1890s. The episode gave credence to the company's two-pronged approach that coupled a tough antiunion stance with paternalistic care for its workers and the City of Kellogg. Though Bunker's sway over its workforce and the community of Kellogg waned a good deal with the replacement in 1942 of the Wardner Industrial Union with the International Union of Mine, Mill and Smelter Workers (IUMMSW, or Mine-Mill), the company's policy of local largesse nonetheless paid large dividends over its long career. Despite the strenuous efforts of union organizers throughout much of the 1930s, Bunker workers stuck with their company union for nearly a decade after it became legal for them to form a genuine trade union.[95] Still, as Mine-Mill's Local Number 18 during the districtwide strike of 1949, Bunker Hill workers garnered the odium of their fellow Coeur d'Alenes unionists with their vote to settle—the "'49 sellout" that broke the strike's back.[96] The watershed strike of 1960, lasting over seven months, deeply divided the community of Kellogg. Charges of communism leveled at Mine-Mill were central to the struggle that ended with the replacement of the IUMMSW with the Northwest Metal Workers Union (NWMW), which Mine-Mill contended was a company union. While it does not appear that the NWMW was in fact a company union, Bunker Hill certainly preferred dealing with it rather than the more militant Mine-Mill, and must have been pleased with the strike's outcome. The outcome of the strike of 1960 represented yet another of many labor victories for "Uncle Bunker" in the wake of the implementation of its paternalistic policies toward its workforce and Kellogg in the early years of the century.

Post–World War I Policies

Following the war, Bunker rehired all of its returning veterans at the same positions or higher ones. Seventeen of the 289 Bunker Hill workers who served the United States during the war did not return. The company memorialized these fallen warriors by placing their names on a bronze tablet on the front wall of its Kellogg offices.[97]

In the mid-1920s, as part of the ongoing effort to, in the language of Bradley and Easton, reward "worthy employees," and to strengthen bonds between Bunker and workers and the community, the company began offering 6 percent preferred stock to employees. The stock paid generous returns—6 percent guaranteed annual interest, plus dividends. Employees could pay for stock either directly or via monthly payroll deductions.[98] Within just over a month the entire stock issue, three thousand shares, had sold out. Believing that a number of "very worthy employees" had been unable to take advantage of the stock offer in the short time it had been available, Easton suggested to Bradley that another, rationed stock offering be issued.[99] Bradley agreed and the company began to offer up to two hundred shares of the preferred stock per month, giving priority to employees who had not previously made purchases.[100]

The Miners' Picnic

In addition to Bunker Hill's direct infusion of cash into the Wardner Industrial Union's coffers and the artificially low utility and rental charges provided to the WIU's offices at the Kellogg YMCA, the company also provided the union long-standing support for its annual August Miners' Picnic. This Kellogg-based extravaganza, held from 1904 through the early 1950s,[101] was a big, multiday, out-

FIGURE 1.2. Undated photo, likely from a Miners' Picnic (S8B-007-6, Barnard-Stockbridge Photograph Collections, Special Collections and Archives, University of Idaho).

door celebration that attracted visitors from across the inland Northwest. In 1937 an estimated ten thousand people crowded Kellogg streets.[102] The Miners' Picnic featured parades and dances, sporting events such as baseball games and boxing matches, theatrical and musical performances, children's races, airplane shows and parachuting demonstrations, mining events such as mucking and drilling contests, and a profusion of jackasses.[103] Political speeches were not uncommon. Idaho governors routinely appeared at the event, and the state's most renowned politician, Senator William Borah, spoke in 1927. For a number of years, Bunker Hill donated the equivalent of a day's worth of company payroll to support the Miners' Picnic.[104]

The Hecla-Bunker Hill Scholarship

In the 1950s Bunker embarked on new community programs while continuing to support long-standing ones. In 1951, in conjunction with the Hecla Mining Company, Bunker's ownership partner in the Star Mine and the zinc refinery, the company launched the Hecla-Bunker Hill scholarship with an endowment-establishing gift of $100,000.[105] The scholarship fund originally was intended to support only the children of Hecla and Bunker Hill employees who wanted to attend Idaho colleges and universities, but soon was broadened to include other local students.[106] Annual scholarships generally totaled between $5,500 and $10,000 and benefited thirty to forty-five students. Through 1977, the scholarship fund disbursed $165,399.41 to college students, with nearly all the money going to young people from the Coeur d'Alenes.[107]

In addition to the scholarship fund, Bunker continued to offer what it referred to as "working scholarships" for local young people. For decades the company made a practice of offering summer employment to local youths. This policy both benefited young locals and provided the company with needed manpower during the summer months when many regular employees took time off to engage in fishing, camping, travel, and other recreational pursuits. According to the company's in-house publication, the *Bunker Hill Reporter*, the summer of 1959, for instance, witnessed that "86 college students and 18 high school students are being employed in Kellogg operations. Of these 104 students all but 13 are from the Kellogg area.... In past years the employment of college and high school students has been as high as 150.... Besides earning a thousand dollars or better... the students fill vacation-depleted crews."[108]

In January 1958 Bunker Hill, with support from the United Crusade of Shoshone County, launched the Idaho Free Ski School.[109] Utilizing the alpine ski facilities at Lookout Pass, twenty-three miles east of Kellogg on the Idaho-Montana border, the company arranged free transportation and ski lessons for six hundred area youths.[110] Through its weekend programs during its years of operation, the Free Ski School enabled thousands of children from the Coeur d'Alenes to ski. Many of these young skiers went on to race competitively, and at least three Lookout Pass ski racers made it to the Olympics.[111]

The Boy Scouts

Over the years, Bunker Hill also gave strong support to the Boy Scouts of America, and particularly to Scouting activities in the Coeur d'Alenes. Many company employees worked with the Scouts, including Stanly Easton, who served as a volunteer Scout leader for forty-three years and was one of the founders of the Boy Scout Council in northern Idaho.[112] Camp Easton, a 383-acre Scout camp on the eastern shores of Lake Coeur d'Alene that annually hosts over 1,300 Scouts, honors the former Bunker Hill president.[113] In 1961 the company began permitting employees serving as Scout Masters to take annual one-week paid leaves of absence so that they could accompany their troops to summer camp.[114] In 1967 the Boy Scouts Twelfth World Jamboree was held sixty-five miles from Kellogg, at Farragut State Park on Lake Pend Oreille. The event drew twelve thousand Scouts and Scout leaders and over eighty-five thousand visitors from all around the world to northern Idaho. Bunker Hill played a significant role in orchestrating the event. Among other things, the company and its employees set up many of the exhibits, donated ore to salt the ground for a "dig" by Scouts, and took about three hundred Scouts on a tour of the company's plants.[115]

Bunker also continued its longtime support for the Kellogg YMCA. At the time of the club's fiftieth anniversary in 1961, the company had added heat to its indoor pool and was offering recreational opportunities that included weightlifting, basketball, volleyball, badminton, and trampoline, in addition to its many youth clubs, summer sports leagues, and night school for adults. The Y claimed nearly two thousand members.[116] The night school remained a thriving concern in the 1960s, boasting an enrollment of 219 in 1966.[117] Some credited Kellogg High School's success in boys' basketball—four state championships between 1955 and 1964—to the fact that the YMCA allowed the team's "gym rats" extensive access to its basketball court.[118] As of 1966 the company had given over $800,000 to the organization. The service of company executives, including Stanly Easton, John D. Bradley, and Wallace Woolf on the YMCA's board of directors in the 1950s, and the extensive coverage granted the organization in the *Bunker Hill Reporter,* also testifies to the ongoing significance of the Kellogg YMCA to the company.[119]

Economic Impacts

Bunker Hill's philanthropic contributions in Kellogg and Shoshone County, though significant, constituted but a fraction of the company's total economic impact on the area. Though the Coeur d'Alenes long boasted a number of large mines, for much of the twentieth century Bunker Hill possessed both the district's largest lead mine and its only lead smelter and zinc refinery. One of the nation's vertically integrated mining, concentrating, and smelting operations, with the capacity to locate and extract raw ores and transmute them into valuable refined metals, Bunker Hill claimed by far the largest workforce, payroll, and assemblage of assets in northern Idaho and was the second-largest employer in the state. The company also boasted the highest revenues and paid the greatest share of taxes in the mining district. In 1951, for example, of an estimated four thousand workers employed in the mining industry in the district, roughly half worked for Bunker.[120]

During the prosperous mid-1950s, Bunker Hill's success, along with that of its fellow mining companies in the Coeur d'Alenes, provided the foundation for a bustling local economy.[121] The 1954 US Census of Business showed that Kellogg led Shoshone County in most retail categories, boasting 21 of the 54 food stores; 27 of 85 restaurants and bars; 10 of 16 car dealerships; 7 of 22 gas stations; 5 of 13 lumber, building equipment, hardware, and farm equipment stores; and 3 of 9 drugstores.[122] The mining county's 7,330 families averaged $4,850 in disposable income for the 1955–1956 business year, or $241 more than the average Idaho family.[123] The trend toward relatively high wages in the Coeur d'Alenes continued until the shutdowns of Bunker Hill and the Sunshine Mine in 1982.[124] In 1975, a year when half of Idaho's $260 million in mining sales came from Shoshone County, mining remained the state's third-highest-paying industry, and the county ranked fourth of Idaho's forty-four counties, with $9,639 in average annual wages.[125]

However viewed, Bunker Hill's impact upon the economy of both Kellogg and a large surrounding region was undeniably significant. Between 1942 and 1981 Bunker Hill annually employed between 1,233 and 3,151 workers and issued between $3,269,149 and approximately $50 million in payroll.[126] In 1961, when Bunker's annual payroll was roughly $13 million, company president C. E. Schwab estimated that Bunker's total cash payments in Shoshone County would be about $19 million for the year.[127] In a talk before the Spokane North Side Rotary Club in 1961, Bunker's manager of metallurgy, A. Y. Bethune, assessed the importance of his company to the economy of Spokane and the Inland Empire, the sizable region, centered on Spokane, that encompasses significant portions of eastern Washington, northern Idaho, and western Montana. Bethune noted Bunker Hill's status as a basic industry—a fundamental economic activity that supports communications, transportation, entertainment, and other service industries. In addition to its substantial payroll and tax payments, Bethune pointed to Bunker's annual purchase of roughly $6 million in operating supplies, over 2.3 million dollars' worth of utilities, the 7 million dollars' worth of railroad freight the company generated annually, much of it traveling on the Union Pacific Railroad, and the over $18 million Bunker paid mines the previous year for their concentrates.[128] Bethune estimated Bunker's value to the Inland Empire at "a tidy $47,000,000 per year." Bethune suggested to the Spokaneites that "no matter what business or profession you're in, chances are whatever affects the Bunker Hill Company, sooner or later affects you for better or for worse. . . . It's to your advantage that we stay strong, vigorous, successful and solvent, that we may contribute our full share to the economy of the region."[129] Though he did not mention it in his address, Bethune also might have noted that a high percentage of Bunker's shareholders—26 percent in 1959, controlling about 50 percent of company stock—lived in the Inland Empire, yet another bond between the company and the region.[130]

While the Bunker Hill payroll (averaging roughly $1.5 million a month, distributed among roughly two thousand Kellogg employees over 1945–1981) undoubtedly did much to propel the local economy, the company's tax payments also contributed to the economy while providing major support for local government services. By the late 1950s Bunker's local property tax payments were averaging about $900,000 annually, and, due principally to the presence of much of the com-

FIGURE 1.3. "Each a Perfect Product"—undated photo of a Bunker Hill parade float.

pany's plant and equipment within its city limits, Kellogg claimed the highest per capita property tax rate of any Idaho city.[131] By 1975, when Bunker's property tax payments stood at approximately $1.36 million, the company's payments represented 22 percent of Shoshone County's total tax revenues, 51 percent of the City of Kellogg's, 41 percent of those for School District 391, 49 percent of those for County Fire District Number 2, 39 percent of West Shoshone County Hospital's, and 22 percent of those for the City of Smelterville. In addition, Bunker provided over $1.3 million in matching Social Security payments for its employees, over $1.6 million in state taxes, and roughly $10 million in federal income taxes.[132]

Gulf Takes Over

On July 26, 1965, for the first time in Bunker Hill's seventy-eight-year history, the company's stock was made available on the New York Stock Exchange.[133] Less than three years later, the historic mining and smelting company fell victim to a hostile takeover by the far smaller Gulf Resources and Chemical Corporation. The

Houston-based Gulf, a company mostly known for its lithium assets in Mexico, utilized a large Wall Street loan to purchase a controlling portion of Bunker stock. On June 1, 1968, Bunker became a wholly owned subsidiary of the Texas firm. No longer run from offices in Kellogg, employees now would have to answer to officials on the twenty-first and twenty-sixth floors of Houston's Tenneco Building. Rather than taking a subservient position in the Gulf-owned company, popular Bunker president C. E. Schwab, having failed in his efforts to fend off the takeover, resigned.[134]

It is fair to say that Gulf was never accepted in Kellogg or the wider Coeur d'Alenes region in the way that Bunker Hill had been. The Houston company continued the Bunker tradition of charitable sponsorship in the Kellogg area—for instance, carrying on the Hecla-Bunker scholarships; sponsoring local television coverage of the annual Kellogg-Wallace high school football game; supporting the Kellogg YMCA; purchasing the local ski hill, Jackass Bowl, when it ran into financial difficulties; and contributing to projects of the Kellogg Chamber of Commerce.[135] Yet no one ever referred to the company as "Uncle Gulf" or felt for Gulf president Robert Allen the respect and affection they had had for Schwab or Easton. Despite their largesse, Gulf and its officials were commonly referred to as "Texas bandits" in Kellogg, where civic leaders tended to view the Gulf takeover as "a bad day for the homefolk."[136]

Although Kellogg was not a company town in the strictest sense, the small city nonetheless was shaped and predominated by the Bunker Hill Company (and then Gulf) for over ninety years—the lion's share of its history. The disappearance of Uncle Bunker from the local scene following the company's shutdown in February 1982 left the community with a gaping economic hole to fill and a profound identity crisis. Even for its local critics, the company had, like the sun, simply always been there.

Kellogg's history and legacy, however, do not consist simply of its record as a highly productive, working-class, company-dominated town—a town long known, in the words of the *Kellogg Evening News*'s masthead, as "The Smelter City of the Coeur d'Alenes," home to the "World's Largest Lead-Silver Mine." The small city and its corporate sponsor left another, at least equally significant legacy that constituted the Mr. Hyde face of their productive activities. Bunker Hill and the small city it built and maintained polluted on an epic scale, fouling water and air and endangering human health to such an extent that it would take decades to remedy much of the damage. Some of their harm to the environment, however, likely will not be undone for centuries or more. The long-known, but often overlooked, substance at the heart of both the economic gain and the environmental destruction of Kellogg and rest of the Coeur d'Alenes is lead.

CHAPTER 2

THE USEFUL METAL

A Brief History of Lead

Lead lacks the prestige and value of its cousins gold and platinum, but the heavy blue-gray metal nonetheless has played a significant role in the development of modern civilization, particularly since the advent of the industrial revolution. For much of the nineteenth century, lead was used primarily for paints and other protective coatings, water lines and pipes, ammunition, ceramic glazes, pewter, leaded glass and crystal, and burial vault liners.[1] The advent of the electrical and automobile age, however, in the late nineteenth and early twentieth centuries created new uses for the durable metal—dubbed "the useful metal" by the lead industry—that dramatically increased its consumption and production. Lead was needed to cover electrical and telephone cables, as a bearing metal, for caulking and solders, and as type metal.[2] Following World War I the rocketing expansion of sales of automobiles with starting-lighting-ignition (SLI) lead-acid storage batteries and terne metal gas tanks again vastly increased demand for the useful metal.[3] While demand for tetraethyl lead (TEL), used as an antiknock additive in "leaded gasoline," did not really take off until World War II, by 1955 America's gas-guzzling, tail-finned cars were consuming nearly two hundred thousand tons of the metal in fuel each year.[4] Postwar prosperity that brought booms in home construction and auto sales, along with a host of consumer goods, also boosted enhanced consumption of "the useful metal." By the early 1970s a typical new American car contained fifty pounds of lead.[5]

Lead's unique combination of important qualities, along with its generally low price, virtually guaranteed its multifarious and heavy employment in industrial societies. Among the metal's more significant qualities are its resistance to corrosion, its malleability and density, and its unique chemical properties. Lead's corrosion and atmospheric resistance was responsible for its use in industrial containers for chemicals such as sulfuric acid and also for its presence in white lead house paints and in red lead coatings for iron and steel.[6] The blue-gray metal's softness and malleability, coupled with its resistance to corrosion, contributed to the use of lead as sheathing for power and telephone lines and its use as solder. Lead's density, availability, and low cost encouraged its use as ballast in ships' hulls and as balance weights in tires. These qualities also granted the mineral ballistic properties that made it a natural choice for use in bullets. In addition to the metal's employment in X-ray protection, these qualities also led to its use as a counterweight in ailerons and rudders that added ease of control and prevented flutter in airplanes. Finally, the metal's unique chemical properties made possible its use in storage batteries, along with its employment as an antiknocking agent in gasoline.[7]

Although it has found many uses in modern industry, lead is hardly a newcomer to humankind. Humans evolved in the presence of minute traces of lead, a mineral that makes up less than 1 percent of the Earth's crust.[8] The metal's use can be traced back at least six thousand years. A lead figure found near the Dardanelles Straits at the site of the ancient city of Abydos can be dated to approximately 3000 BCE.[9] Unsmelted lead ore, known as galena, was used in eye cosmetics as early as 4000 BCE. By the fifth century BCE, the ancient Athenians mined galena at the nearby Laurion mines and produced up to ten tons of lead and silver each year.[10] But while ancient Greece may have used a good deal of lead, ancient Rome made the gray metal fundamental to their civilization.

By 50 BCE, lead had become an integral part of Roman life. Among many other things, the metal could be found in wine vessels, tombs, tooth fillings, bathtubs, and most ubiquitously, Rome's extensive hydroengineering systems.[11] These systems employed massive quantities of lead to bring water from the mountains to the plumbing systems of Rome and other cities.[12] The Greeks and Romans, however, were far from the only early civilizations to use the versatile metal. Egyptian pharaohs and Assyrian kings inscribed messages on lead sheets, while the ancient Chinese wrote on bamboo with white lead ink. The ancients also associated lead with the gods. While ancient Egyptians associated lead with Osiris, Greeks and Romans linked the metal to Cronus/Saturn.[13]

While lead is deeply intertwined with the history of human achievement, there is another side to the human relationship with "the useful metal." For as long as people have worked with lead, they have experienced its toxic properties. The blue-gray metal is, in fact, a potent neurotoxin that accumulates in the body's bones and soft tissues. Among the metal's more significant health effects are nerve damage, anemia, kidney dysfunction, hypertension, reproductive damage, and brain damage. In its chronic form, lead poisoning can result in slowed reaction time, lead neuropathy (i.e., wrist or foot drop), kidney failure, depression, and short-term memory loss.[14] Severe lead poisoning may lead to delirium, coma, encephalopathy (a severe condition whose symptoms include brain swelling), or death.[15] Lead harms—and in higher doses kills—by disrupting the intricate machinery of red blood cells. Substituting for necessary metals such as zinc and iron in cells, lead wreaks havoc, preventing blood from creating heme, blocking nerve conduction, and interfering with mitochondria.[16] Although potentially harmful to adults, lead poses the greatest threat to young children and to fetuses, due to the fact that their fast-growing bodies absorb the toxin so rapidly.[17]

The realization that lead is a potent poison is far from a recent phenomenon. In the second century BCE, Greek poet and physician Nicander described most of the classic symptoms of saturnism, or lead poisoning—including hallucinations and paralysis—and recommended several purgative treatments. Until the nineteenth century, the most common cause of lead poisoning was probably the drinking of wine or cider stored in leaded containers. This malady appears to have been widespread in ancient Greece and Rome; some scholars go so far as to blame the fall of Rome on wine-induced saturnism. The discovery that lead retarded fermentation and made wines taste sweeter persuaded vintners to add the mineral. In 802 CE, Charlemagne is reputed to have called for a ban of leaded wines. During the Renaissance, the German city of Ulm executed those convicted of adulterating wine with lead. Severe bouts of colic that wracked Poitiers, France, and Devonshire, England, in the seventeenth and eighteenth centuries were later found to be products of, respectively, lead-tainted wine and cider. In 1723 the Massachusetts Bay Colony banned the use of "leaden heads or worms" in the distillation of rum or other spirits.[18] Benjamin Franklin, who observed that printers develop "dangles and gripe" (wrist drop and colic) from lead poisoning, commented in 1786, "The opinion of this mischievous effect from lead is at least above sixty years old; and you will observe with concern how long a useful truth will be known and exist before it is generally received and practiced upon."[19]

In the nineteenth century, with the rise of industrialism, interest in the effects of lead on workers grew. Louis Tanquerel de Planches published the pioneering study on occupational lead poisoning in 1839. Based upon his examination of the case histories of over 1,200 patients at La Charité Hospital in Paris, Tanquerel's work detailed the neurotoxic effects of lead and showed that lead compounds, especially pigments, posed great occupational threats.[20] While Tanquerel's report prompted major voluntary improvements in the hygienic conditions in many French lead pigment factories, reforms in other countries did not come as quickly.

More than any other figure, Alice Hamilton focused attention in the United States on the dangers of occupational lead poisoning. Hamilton, a medical doctor by training, was inspired by her long experience living and working at Chicago's Hull House to try to ameliorate the conditions of life of the urban poor.[21] Reading

British physician Thomas Oliver's work encouraged Hamilton to pioneer the field of occupational safety in the United States.[22] In the first decade of the twentieth century, however, America lagged far behind France, Germany, and Great Britain in its protection of industrial workers from occupational illness and other job hazards. In 1907, the Illinois Commission on Workingmen's Compensation enlisted Hamilton to spearhead a study on occupational disease in the state. While the study examined workplace exposure to a number of known toxins, Hamilton focused her own work on lead. Hamilton and her staff visited 304 lead industry plants in 1910–1911. The commission's findings, published in 1911, uncovered abundant evidence of plumbism in Illinois factories. Following the Hamilton study, it was no longer possible to claim seriously that American industry had no lead problem.[23]

The Deadly Power of Lead

In the mid-1920s the useful metal put on an unforgettable display of its versatility and potency as a workplace toxin. In December 1921, Thomas Midgley Jr., chief chemist with the Dayton, Ohio–based General Motors Research Corporation, discovered tetraethyl lead's benefit to gasoline-powered engines as an antiknock agent. Unlike lead oxides such as those found in white lead paints, however, tetraethyl lead is fat-soluble and highly toxic to the nervous system when absorbed through human skin.[24] Indeed, the US Department of War tested TEL as a possible nerve agent during World War I, but found it less effective than mustard gas for combat purposes.[25]

In October 1924 events at Standard Oil's Bayway tetraethyl lead production facility in Elizabeth, New Jersey, amply demonstrated TEL's lethal tendencies. Although DuPont was licensed to manufacture the bulk of the new Ethyl Gasoline Corporation's patented TEL, Standard Oil of New Jersey also was permitted to manufacture the substance at its small Bayway plant.[26] On October 23, 1924, things began to go terribly wrong at Bayway. On that day, Ernest Oelgert, a worker at the facility (dubbed "the loony gas building" by workers), began to experience hallucinations, complaining to coworkers that people were following him. The following day, running through the factory, he shouted that there were "three coming at me at once." Caught and taken to a nearby hospital, Oelgert died in convulsions the next day. Over the next week, four of Oelmgert's co-workers died in similarly gruesome fashions. Thirty-five of the "loony gas building" workers who survived nonetheless suffered tremors, hallucinations, severe palsies, and other serious neurological symptoms of organic lead poisoning.[27] In all, forty of Bayway's forty-nine workers that fatal October experienced either severe neurological illness or death.[28]

Journalists soon discovered that the Bayway deaths were not the first to be caused by the manufacture of tetraethyl lead. The previous April, two workers at GM's experimental facility in Dayton had succumbed, while more than fifty others had contracted serious lead poisoning.[29] At DuPont's Deepwater plant in southern New Jersey, at least six died and over three hundred were poisoned before the Bayway tragedy.[30] Of the three TEL manufacturing facilities, Deepwater was by far the largest. Designed to produce 1,300 pounds per day, the plant was rushed into service before proper ventilation was installed and before a safe manufacture process had been established.[31] Like Bayway, Deepwater received a moniker from workers. It was dubbed "the House of Butterflies," for the creatures that the poisoned workers saw flying around them in the facility.[32]

Within days of the last of the Bayway worker deaths, New York City, Philadelphia, Pittsburgh, and New Jersey banned the use of gasoline containing the lead additive.[33] In late April 1925, H. S. Cumming, the surgeon general of the Public Health Service, announced that he was calling together experts from business, labor, and public health to assess the tetraethyl lead situation.[34] On May 20, eighty-seven participants convened in Washington, DC, for the surgeon general's conference.[35] At the conference TEL defenders offered three principal arguments: first, that the product was essential to the nation's industrial progress; second, that any innovation entailed certain risks; third, that the failure of TEL workers to exercise proper caution had caused their illnesses and deaths.[36] Those opposing leaded gasoline presented a fundamentally different viewpoint. They argued against the risk of introducing a powerful cumulative toxin into the environment on such a large scale.

Before closing, the conference called upon the surgeon general to organize a committee of the nation's top public health scientists to conduct a study of leaded gasoline. The surgeon general assembled an all-star committee and requested that it make its report in a timely fashion. The committee designed a short-term and rather limited study of 252 Dayton, Ohio, chauffeurs, garage workers, and filling station attendants, some of whom came into contact with TEL while others did not. The committee's report, released in January 1926, concluded that "there are at present no good grounds for prohibiting the use of ethyl gasoline . . . provided that its distribution and use are controlled by proper regulations."[37] Although the committee noted "a greater storage of lead in the bodies of those exposed to ethyl gasoline," and advised continued investigation of the health effects of leaded gasoline, the TEL industry reacted simply as if it had been given a clean bill of health. On June 1, 1926, the Ethyl Gasoline Corporation reintroduced its only product to the nation. "Ethyl Is Back," read signs in filling stations.[38] No public studies of the health and environmental effects of leaded gasoline would occur until the mid-1960s, a period in which all scientific studies of tetraethyl lead were conducted by the Ethyl Corporation or scientists the company employed.[39]

There was, however, a lead health problem that the occupational health scientists' work did not begin to address. In the 1920s the US medical community was only beginning to recognize the nature and the scope of the problem, and it would be decades before the public at large became conscious of "the silent epidemic." Childhood lead poisoning, caused by eating leaded paint chips, killed hundreds of children every year in America's cities. Fatal cases of childhood lead poisoning, however, were commonly misdiagnosed as tubercular (TB) meningitis and other diseases.[40] For a variety of reasons, the foremost among them the preoccupation with germ-based causes of disease, American doctors of the late nineteenth and early twentieth centuries were blind to the presence of environmentally caused diseases such as childhood lead poisoning.[41]

White lead-based paints retained the leading position in the paint industry until the 1940s when they were supplanted by titanium dioxide pigments.[42] A significant portion of house painters were fervent "white leaders," believing that "pure" white lead provided superior durability and overall quality. While many European countries banned the interior use of lead paint following World War I, until the 1970s lead-based paints were used commonly on interior as well as exterior walls in the United States.[43]

From the late 1920s through the late 1940s, childhood lead poisoning was defined in the United States largely as a disease of poverty that could be safely ignored by the bulk of the population. Sensational cases such as the Depression-era "battery burnings," in which unfortunate individuals died after burning inexpensive wooden battery casings to heat their homes, served to solidify this definition.[44] In the 1950s and 1960s, as the health departments of major cities became more aggressive in their pursuit of childhood lead poisoning, the disease was rebranded as a problem of race, a "ghetto disease," due to the fact that African Americans were far more likely than any other group to suffer from it. Although a number of prominent childhood lead poisoning researchers objected, many in the medical community defined pica, or the tendency of children to eat nonfood items such as lead paint chips, as the race-based behavior that lay at the root of childhood lead poisoning. Going further, some equated pica with the "exotic" cultural practices of dirt eating acquired during the days of sharecropping and slavery in the American South, and blamed contemporary African American mothers for not keeping their children from eating paint chips.[45] This reasoning conveniently segregated the disease while also laying blame squarely on its victims.

By the 1960s the enhanced ability of US doctors to correctly diagnose childhood lead poisoning, along with the use of powerful chelating agents—chemicals that bind with and neutralize lead in the body—such as British anti-Lewisite (BAL), significantly lowered the death rate from the disease.[46] In New York City, for example, 80 cases of lead poisoning were reported in 1954 with 12 deaths, whereas in 1964 there were 509 cases and 7 fatalities.[47]

In 1977, the federal Consumer Product Safety Commission effectively banned lead paint from the US market.[48] While a public health hazard undoubtedly lingers in the form of the estimated twenty-four million US homes still containing lead paint, major strides have been made in the battle against childhood lead poisoning over the past forty years.[49] Childhood deaths from lead poisoning have gone from commonplace to almost unheard-of, while at the same time, and principally due to the phaseout of leaded gasoline, median blood lead levels in the United States have decreased by nearly an order of magnitude.[50]

Researchers, however, continued to unearth evidence of bodily harm at increasingly low levels of lead exposure. In 1971 Herbert Needleman, a physician and psychiatrist, studied the lead content in the baby teeth of 761 Philadelphia schoolchildren and showed a clear association between lead absorption, environmental lead pollution, and poor housing. In 1979, Needleman and his associates published a pathbreaking paper attributing diminished IQs and school performance to subclinical lead exposure. The study, conducted in Boston, again relied upon lead levels in the baby teeth of schoolchildren. The 273 children studied were mostly healthy; only ten showed symptoms of lead poisoning and had blood lead levels above 40 μg/dl.[51] Needleman concluded that a strong correlation existed between higher levels of tooth lead and lower IQ scores and poorer school performance.[52]

In response to the growing body of research showing problematic health effects from low-level lead ingestion, the medical community has repeatedly lowered its threshold levels for "safe lead exposure." Following significant reductions of the standard in the 1970s, in 1985 the federal Centers for Disease Control (CDC) dropped the threshold for pediatric concern for blood lead from 30 to 25 μg/dl. In 1991, the CDC again lowered the level, this time to 10 μg/dl, and called for a plan to eliminate childhood lead poisoning by 2010.[53] Unlike selenium and other minerals that are necessary to proper body functioning in small doses but toxic in large ones, it now appears that no amount of lead is good for the body and that even small doses cause harm.

The Phaseout of Leaded Gasoline in the United States

The initial death blow for leaded gasoline came from a most unlikely source: General Motors. The catalytic converter, an antipollution device that converts carbon monoxide to carbon dioxide and water, was developed in the late 1960s. From the perspective of the leaded gasoline industry, however, there was a big problem with the device. It was rendered inoperable by even trace amounts of lead. In January 1970 GM president Ed Cole announced the company's intention to meet the conditions of pending clean-air legislation with catalytic converters beginning in 1974.[54]

At the same time, the new federal Environmental Protection Agency (EPA) grappled with its mandate under the Clean Air Act of 1970 to regulate serious air pollutants such as lead. In 1972 the EPA proposed two separate regulations covering leaded gasoline. The first required that, in order to protect catalytic converters, all gasoline producers offer at least one grade of unleaded gasoline by 1975. The second regulation, designed to protect human health, required gasoline refiners to halve the amount of lead in the nation's gasoline pool within four years' time.[55] In 1978, following a lengthy battle between government and lead-industry-funded scientists, the EPA promulgated a national ambient air lead standard of 1.5 μg/dl.[56] The new air lead standard would affect a number of American communities whose air was lead-saturated, including lead smelter towns such as El Paso, Texas, and Kellogg, Idaho; and smog-filled cities such as Los Angeles. The lead industry challenged the new standard in court but ultimately lost its battle before the DC Court of Appeals in 1980. In 1985, the EPA announced plans to speed the phaseout of leaded gasoline, and ten years later Congress completely banned its sale.

By the mid-1980s the old toxicological paradigm for lead health had been thoroughly overturned. In its place reigned a scientific consensus that viewed almost any lead exposure to children as potentially harmful. This new consensus, the product of a nearly classic case of what the historian of science Thomas Kuhn has labeled a "paradigm shift," relied upon studies demonstrating damage to mental functioning, nerve conduction, and other bodily processes at ever-lower levels of lead exposure.[57] Over the past forty-five years, the momentum to eliminate potentially hazardous lead-based products from the US market has remained remarkably consistent. In 1986 the EPA reduced the permissible lead contamination of drinking water by a factor of ten, prompting the passage of federal laws that sharply curtailed lead in solder, plumbing fixtures, and drinking fountains.[58]

The elimination of the lion's share of readily bioavailable lead from the US environment has had a profound impact upon the overall health of Americans. By far the most significant step in terms of lowering bodily lead burdens has been the elimination of leaded gasoline. The Second National Health and Nutrition Survey (NHANES II), conducted by the CDC between 1976 and 1980, showed that blood lead levels had dropped by nearly 40 percent over the period, from 14.6 to 9.2 μg/dl, closely paralleling the decline in sales of leaded gasoline.[59] By 1991, when leaded gasoline had been nearly eliminated from the US marketplace, average blood lead levels had fallen to 2.8 μg/dl, just over one-fifth of where they had stood fifteen years earlier.[60]

In the early decades of the twenty-first century, lead, the once-ubiquitous, seemingly indispensable metal of modern industrial civilization, has been largely swept into the corners of American life, though from time to time the useful metal still rears its head as a potent menace to public health, as it has done recently in Flint, Michigan, and East Chicago, Indiana. While US and international lead consumption have continued to expand slowly, its uses have been dramatically curtailed. In the United States, the metal is no longer put into solder, pipes, wire sheathing, gasoline, or paint.[61] In the United States, since the mid-1980s the percentage of total lead consumption devoted to lead-storage batteries has increased dramatically. By the early 2000s, batteries accounted for nearly 90 percent of all the lead consumed in the nation.[62] With the rapid growth of car purchases in major developing economies like China's and India's, the near-term future of the effective, inexpensive lead-acid battery seems relatively secure. The popularity of the lead-acid battery, in turn, secures the future of the global lead mining, smelting, and recycling industries.

China, the economic juggernaut of the past three decades, has emerged as the world's leader in lead smelting. It is not surprising that environmental and human health problems have arisen in tandem with the rise of China's lead smelting industry. In 2009, in the city of Jiyuan, home to China's largest lead smelter, about fifteen thousand workers were forced to relocate after more than one thousand children exhibited excessive levels of lead in their blood.[63] In 2007, in an eerie reprise of a public health problem long considered resolved, over ten million toys sold by Mattel Corporation were recalled because they contained lead paint. The Chinese manufacturers who assembled the toys for Mattel used the toxic paint because of its relative cheapness.[64]

Recent incidents involving the Chinese lead industry serve to illustrate the global tendency to shift highly toxic industries such as lead smelting and textile dye manufacture from the developed to the developing world. While in the United States and other developed countries lead health issues may have assumed a diminished importance—though Flint's and East Chicago's problems highlight the risk that lead can still pose to human health—the imbrication of the still-useful metal in the global economy ensures that people will continue to suffer from exposure to the substance. That suffering, however, has shifted largely from economically advanced nations that have taken steps to protect workers and consumers to less-developed countries loath to impose strict regulations for fear of losing their lead industries.

In northern Idaho's Coeur d'Alenes, home to intensive and extremely profitable lead and silver mining for over a hundred years and smelting for over sixty, the cleanup continues, more than thirty years after the Kellogg smelter's last shift call. Although the federal EPA Superfund cleanup already has exhausted three decades and hundreds of millions of dollars, there is no end in sight. It is evident that, across the globe, the bill for environmental damages from lead mining, smelting, and use will be coming due for decades to come.

CHAPTER 3

LEAD CREEK

It begins as a trickle high in the Bitterroot Mountains just west of Lookout Pass.[1] Here, a stone's throw from the Montana-Idaho border, the South Fork of the Coeur d'Alene River runs cold and clear. Healthy populations of native cutthroat trout swim about. West of the town of Mullan, Idaho, about ten miles downstream from the headwaters of the South Fork, however, the picture begins to change. The stream's water becomes more turbid; fewer fish are to be found. By the time the South Fork reaches the town of Wallace, another eight miles downstream, although its water is still refreshingly cold, there are no fish. What has happened, one wonders? Where did the fish go?

Simply put, mining happened. From the hills just northeast of Mullan, where the Lucky Friday Mine has operated for nearly seven decades, to the flats west of Kellogg, where the giant Bunker Hill Company smelter once belched out its toxic smoke, lies one of the richest mining and smelting districts in world history. Since the time that gold was discovered near Murray in 1883, and, far more importantly, silver-bearing galena was discovered near Kellogg in 1885, to 1997 (the most recent date for which comprehensive figures were available), the Coeur d'Alene mining district yielded staggering amounts of metals: 34,300 metric tons of silver, 2,870,000 metric tons of zinc, and 7,288,000 metric tons of lead, or 18, 17, and 6 percent of the nation's production of those metals, respectively.[2] The total monetary value of these metals stands in excess of $26 billion in 1997 dollars. In terms of historic minerals production in the United States, the Coeur d'Alene district ranks first in silver, third in lead, and third in zinc. Indeed, the district has produced more silver than any other mining area in world history, even more than the legendary Cerro Rico de Potosi district in Bolivia.[3] But the surrounding landscape and the local human population undoubtedly have paid dearly for these revenues.

Some of the largest, most profitable, and most famous mines in the United States have been located in the Coeur d'Alenes—among them the Lucky Friday,

Golconda, Sunshine, Gold Hunter, Hercules, Hecla, Page, Morning, Star, and Bunker Hill. These mines have employed tens of thousands of workers and paid out many millions of dollars in wages, sustaining local communities that, at their peak in 1950, boosted the Silver Valley's population to 22,500. Though the largest of these concerns, the Bunker Hill Company, possessed the largest lead and zinc mine in the United States, the Bunker Hill Mine, it is but one of eighteen mines in the Coeur d'Alene mining district that has yielded over a million tons of metal.[4] But they and their brethren, the district's ore concentrators and the Bunker Hill smelter, possess another legacy, one of environmental degradation and destruction.

This chapter examines the history of the Coeur d'Alene watershed from the advent of mining along tributaries of the South Fork of the Coeur d'Alene River in the early 1880s through the present. It foregrounds the shifting ecological health of the watershed, and seeks to connect the industrial technologies and processes of the mining district to downstream "changes in the land."

Mines used the South Fork as an industrial sewer until the late 1960s, in keeping with the industry's common practice during the era. This led to profound wreckage of downstream ecologies. The mining companies' dominance of the Coeur d'Alenes' economy and of the region's environment, potently demonstrated by their intense fouling of the watershed, did not go uncontested, however. By the early years of the twentieth century, farmers and ranchers concerned by the despoliation of their irrigation ditches and fields by mine wastes were jousting in court with upstream mining interests who used the South Fork to cheaply transport tailings that otherwise might overwhelm the mining district. By the late 1920s Coeur d'Alene City merchants reliant upon tourism to scenic Lake Coeur d'Alene had joined the fight against the pollution from the mining district. Successfully overcoming their opponents, however, the mining and milling companies' power to pollute the watershed remained supreme, and nearly absolute, from the 1880s through the late 1960s. This power, and its ecological consequences, ultimately deprived downstream farmers, ranchers, sportsmen, and merchants of the ability to use the watershed and its adjacent lands as they wished.

The story of the Coeur d'Alene River watershed cannot be truly understood without a realization of the inherent risks, tradeoffs, and sheer environmental destructiveness of industrial modernity. In 1885, the sparkling Coeur d'Alene River teemed with fish; by the 1930s it ran gray and was nearly devoid of aquatic life. In the interval, of course, a thriving mining, milling, and smelting district had arisen in and around Kellogg. While the district's corporate managers experienced the benefits of industrial development, downstream users primarily felt its risks and costs, and the environment of the Coeur d'Alene basin suffered devastation. An

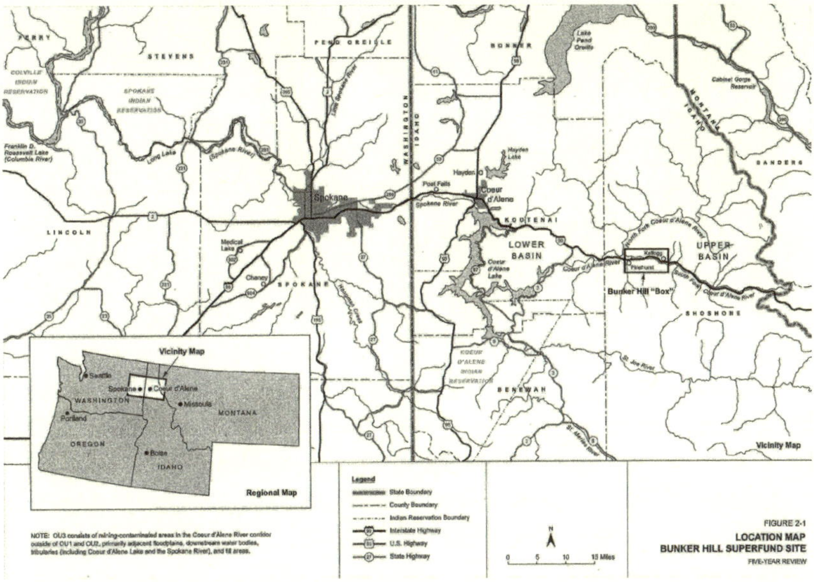

FIGURE 3.1. Location map of Bunker Hill Superfund site (US Environmental Protection Agency).

early 1930s controversy over the fouling of Lake Coeur d'Alene by mine wastes briefly focused the State of Idaho's attention on the ecological health of the drainage basin. This episode offers an interesting window on the competing values and interests that long shaped the basin's ecology.[5]

Early Days on the Coeur d'Alene

Although today, even after decades of cleanup efforts, one is hard-pressed to find living fish in the portion of the South Fork of the Coeur d'Alene that wends its way through the heart of the old mining district, the situation was not always so. Until the arrival of significant numbers of Euro-Americans in the area following the completion of the Mullan Road in the early 1860s, the Silver Valley had been primarily the preserve of the Coeur d'Alene American Indian tribe, who utilized its extensive wild game and fish resources. Jesuit missionaries such as Father Nicolas Point, who arrived in the area to proselytize to the Coeur d'Alenes in the early 1840s, noted the richness and fecundity of its natural landscape. Describing the place where Lake Coeur d'Alene empties into the Spokane River, Point remarked, "The waters are teeming with fish which are caught, until January, by means of a trellised barrier extending from shore to shore."[6] Point also noted a local hunt in which six hundred deer were taken. Discussing the plentitude of game animals roaming over the Coeur d'Alene Indians' landscape, encompassing the Silver Valley and considerable territory to its west, north, and south, Point wrote:

> Is there an abundance of game? Perhaps nowhere does so small an area contain such a variety. Next to the roe deer, these are the most common: the deer, the elk, the mountain lion, the cacajou, the white sheep, the bighorn, the goat, the wolf, the fox, the wildcat, the polecat, the hare, the otter, the weasel, the badger, the mink, the marten, the fisher, the beaver, the muskrat, a large variety of mouse-colored rats, squirrels, field mice, not to mention four or five varieties of bear.... Fish are abundant in the lakes, rivers and small streams.[7]

Commenting on the same area twenty years later as he oversaw the building of the federal road that would bear his name, Captain John Mullan noted that the Mission Flats section of the Coeur d'Alene River, about fifteen miles west of Kellogg, hosted a sufficient supply of fish to provide vital sustenance to a population of three hundred Coeur d'Alene Indians who resided at the Jesuit mission there.[8] In 1883, on the eve of the mineral discoveries that would forever change the region, a resident named Stoll commented that the rivers of the Coeur d'Alene country "teemed with fish."[9] In 1930, Armand Perrenoud, an "old-timer" who had traveled the Coeur d'Alene River by boat in the 1880s, reported that in those days the river possessed remarkable beauty and charm and supported adjacent farms that raised ninety to one hundred bushels of oats and five tons of hay per acre.[10] Writing about the Coeur d'Alene River in her book on the area's once-prolific steamboat traffic, Ruby El Hult noted that in the 1890s, during the heyday of the passenger steamboat era, "the Coeur d'Alene River was ... clear as crystal, deep, with cottonwoods and silver beeches on both banks almost forming an arch overhead."[11] Describing the Coeur d'Alene's fecundity, Hult wrote, "The stream was alive with trout and other fish, and all the early boats featured trout dinners. Just below the Old Mission, at the Big Eddy where the boats turned around, scraps were thrown into the water nearly every day. Most of the fish in the Coeur d'Alene River came to haunt this feast table. They could be seen by the thousands in the clear water and were a source of amazement to the boat passengers."[12]

In the fall of 1883 the discovery of gold along Prichard Creek led to one of the more memorable American gold rushes of the late nineteenth century. Spurred on by an aggressive publicity campaign by the Northern Pacific Railroad, which hoped to people its trains with westward passengers, thousands of would-be prospectors descended upon the remote wilderness outpost during the winter of 1883–1884. Once arrived, the gold hunters quickly cobbled together the towns of Murray and Prichard. The frenzied placer-mining era, however, was short-lived. By

1886 it was clear that all, or nearly all, of the easy-to-pan gold in the Coeur d'Alenes had been scooped away. Though some success still was to be had in acquiring the glittery metal through quartz mining techniques, including the use of a dredge in the early twentieth century, the romantic placer-mining era in the Coeur d'Alenes was largely finished by the late 1880s. The district's future lay in other, less hallowed metals.

In September 1885, Noah Kellogg's jackass made its famous find, ushering in the rush of industrial mining capital, labor, and technology to the Coeur d'Alenes and leading to the creation of the town of Kellogg (as was discussed in chapter 1).

As was common practice in the late nineteenth century and for decades thereafter, Silver Valley mining ventures of all sizes employed the nearest serviceable moving body of water to transport their two types of tailings—*jig tailings*, sand-sized particles that accumulated in large piles along creeks, and *slimes*, very fine-grained, sub-silt-sized tailings that would travel much farther downstream—away from their mining and concentrating operations.[13] In this case, that meant the South Fork of the Coeur d'Alene River and its numerous immediate tributaries. The quantity of such tailings, even in the early years of Silver Valley mining, was sizable. Mineral economist Keith R. Long estimates that from 1886 to 1997, 56 million metric tons of tailings "were disposed of or otherwise lost into the Coeur d'Alene River or its tributaries."[14]

Milling Processes and Environmental Problems

For the first few decades of mining in the Coeur d'Alenes, companies used the stamp and jig table method of separating valuable ores from waste rock. Mechanical stamps crushed the ore and this ore was then sent over the jig tables. As in panning for gold, this method utilized the propensity of heavier ores such as silver and lead to sink to the bottom in water, separating more valuable ores from waste rock. A drawback of the jig table method, however, was its relative inefficiency: between 20 and 40 percent of the lead and silver was lost to the tailings pile. Most tailings from the early era—1885 to roughly 1925—tended to be rather coarse and large. Of the more than one hundred milling operations in the Coeur d'Alenes during this period, the bulk of them located along the upper South Fork and its nearby tributaries, most used flumes to transport their tailings directly from the mill to a large tailings pile. Due to the steepness of valley walls and the narrowness of canyons in the district, flooding and high rates of soil erosion are commonplace. This led to the frequent washing of coarse tailings into the South Fork of the Coeur d'Alene and its tributaries. During the jig table era, however, most tailings, due to their size, generally did not wash more than about ten miles downstream. Long has concluded that, nonetheless, by 1900 mill tailings had reached Lake Coeur d'Alene (thirty-five miles downstream from Kellogg) and had affected as much as twenty-five thousand acres of floodplain along the river.[15]

Beginning around 1910 a new method of concentrating ores came into use in nonferrous mining, including in the Coeur d'Alenes. Gradually gaining acceptance throughout the district, it was almost universally employed by 1930. This method, the flotation process, employed chemical reagents and oils that would bind to the valuable ores in a slurry of water to separate them from the waste rock.[16] Flotation allowed ore-recovery rates to increase to as high as 90 percent, meaning that far fewer tailings would be dumped into the river. But there was a downside. Tailings produced via the flotation process were far finer than those produced using the jig table method. This meant that tailings created under the new process would be carried much farther in suspension than they had been previously, often all the way to Lake Coeur d'Alene, if not farther. While the flotation process helped solve a difficulty for concentrator operators—the accumulation of large tailings piles and bars near their operations that could threaten to impede their milling—it also was instrumental in creating a heightened environmental problem: the widespread diffusion of lead and other toxic minerals throughout the watershed of the South Fork of the Coeur d'Alene River.

Downstream Farmers and Ranchers Fight Back

In 1899, downstream farmers and ranchers began to complain that debris from mining and milling was fouling their land and wreaking havoc upon their crops and livestock, particularly in locations where the Coeur d'Alene River usually flooded in the spring. In 1903, Josiah Hill, a Shoshone County farmer, filed suit against the Standard Mining Company, which operated a mining and milling operation on Canyon Creek above Wallace. This would be the first of a number of environmental cases against the mining concerns of the Coeur d'Alenes by downstream water users. The following year, as the Hill case worked its way through the Idaho courts, sixty-five agriculturalists from Kootenai County (which lies immediately downstream from Shoshone County) filed a pair of class actions against the mining companies. The first case, *McCarthy et al. v. Bunker Hill and Sullivan Mining and Concentrating Co. et al.*, sought an injunction to stop further polluting of the Coeur d'Alene River and its tributaries. The second, *Doty et al. v. Bunker Hill and Sullivan Mining and Concentrating Company Co. et al.*, sought over $1,224,000 in pollution damages. Plaintiffs' testimony in the two class action cases focused primarily upon the same sorts of alleged wrongs: "leaded" soils and plants in farmers' fields, diseased and otherwise damaged livestock, greatly diminished hay yields, dead fish in a now-polluted river, and gray-colored and "leaded" water issuing from the mining district.

Elmer Doty, the lead plaintiff in the second lawsuit, testifying in the *McCarthy* suit in 1905, spoke of dead and sickened cows and horses, describing horses that lost hair and developed severe breathing difficulties and lethargy and cows plagued with severe diarrhea. The farmer described hay fields whose yields diminished from two tons to a quarter ton per acre, the grayish color of the river water that flooded his fields and a grayish dust that afterward was left upon his vegetation, and numerous "bald spots" where no hay would grow in previously flooded fields.[17] The *Doty* plaintiffs contended that widespread deposits upon the lands adjacent to the South Fork and the main stem of the Coeur d'Alene River "contains both lead and zinc in large and poisonous quantities . . . [and] that a large number of horses and cattle died from eating this hay and grass that was grown on these meadows."[18] They argued that the market for the hay they grew along these rivers "was destroyed more than four years ago."[19] The farmers alleged that while the federal Department of Agriculture held that as little as a tenth of a grain of lead in solution per gallon of water (or 17.14 parts per million) was poisonous to livestock, there was twenty-five times that quantity in the waters of the Coeur d'Alene River. Plaintiffs charged that the mining interests had dumped twelve million tons of material into the watershed, and that, not surprisingly, river channels were filling up with sediment. This of course made seasonal flooding more prevalent and more intense. Plaintiffs also noted that while the North Fork of the Coeur d'Alene River, where mining activity was not significant, flowed "pure," the South Fork was a wholly different, and far dirtier, story. Addressing the health of the South Fork itself, plaintiffs stated that in 1898 farmers "noticed that the fish were destroyed on the river, [and] no more fish appeared in that pure stream. Many of them floated down the stream belly up, and no fish have ever inhabited those waters since."[20]

In the initial case filed against the upstream mining interests, Josiah Hill, a Shoshone County farmer, initiated a twelve-thousand-dollar damage suit against the Standard Mining Company, alleging that the company had dumped 550,000 tons of mining debris containing "poisonous substances" in Canyon Creek. This material, Hill contended, then had washed down into the South Fork of the Coeur d'Alene,[21] where it had filled the river's channel to such an extent that it caused a flood that deposited sediment on his land. Hill alleged that this sediment destroyed his crops and the value of his agricultural land, while also contaminating his well water. The company's lawyers responded that the Idaho Constitution granted "preference" to mining operations in the use of water, and for good measure added that the accusers had failed to illustrate clearly Standard's responsibility and negligence in the alleged damages incurred. The district court in Wallace granted the defendants a discharge, but the Idaho Supreme Court consented to hear the case on appeal.

The court, in a historic opinion written by Chief Justice Charles O. Stockslager, reversed the lower court and held that despite their preferential status vis-à-vis water appropriation, mining companies did not enjoy an unlimited right to load stream channels with "debris and poisonous substances to the injury of other users of water."[22] Downstream water users, ruled the court, had "the right to assume the upper appropriator will continue the use of water as he found it, and if any change would damage him in the use of his appropriation, the courts will protect him in his right."[23] In its ruling, the court rejected the flatly utilitarian, economic argument made by the corporation that should the court rule for the plaintiff, the result in Shoshone County well could be "the abandonment of all mining and milling . . . and bankrupting of [its] inhabitants." Stockslager, writing for the court, responded, "Deplorable as this might be—if true—it furnishes no excuse for the Court to shirk its responsibilities." "The law," wrote the chief justice, "protects the lone settler in his rights, let them be ever so meager, as well as the capitalist, the corporation or the individual with its or his millions."[24]

While Hill personally garnered little from the court for his efforts—he was granted only the cost of his appeal—the court's decision nonetheless represented a significant check, at least in theory, on the right of Idaho mining companies to use the state's streambeds as dumping grounds for their tailings. Because many of the mining companies operating in the Coeur d'Alenes were registered as "foreign," out-of-state corporations, however, most of the cases following *Hill* were tried in federal court. In *McCarthy*, concluded in November 1908, plaintiffs' request for an injunction to stop all mining activity in the district was denied. Judge James H. Beatty of Boise, while admitting that the "complainants have suffered injury and may suffer more from [the] alleged causes," nonetheless concluded that the "preponderance of the evidence" in the case seemed to be largely in the mines' favor.[25] Unlike the Idaho Supreme Court, the federal court proved willing to consider the economic value of mining to Shoshone County in rendering its decision, stating that the closing of every mine and mill in the district would paralyze the sole means of occupation of nearly all of its twelve thousand inhabitants. "Any court," wrote Beatty, "must hesitate to act as to bring such a result."[26]

The Doty case also denied relief to the farmers and ranchers. In a judgment in 1910, Federal Judge Frank S. Dietrich, despite finding that evidence showed injury to the agriculturalists, nonetheless held that there was no support for the claim that Bunker Hill was "responsible for the condition of the water . . . it can only be held responsible for its own wrongful acts."[27] Accepting Bunker Hill testimony to the effect that water leaving its impound area was cleaner than the water the company received from upstream users, Dietrich found that there was "evidence insufficient to justify a verdict in favor of the plaintiffs for more than nominal damages."[28]

Commencing in 1901, the Mine Operators Association (MOA),[29] in response to the complaints of downstream agriculturalists, began building relatively crude dams of wood pilings and planks to impound tailings along Canyon Creek and the South Fork of the Coeur d'Alene.[30] The Canyon Creek Dam was constructed in the vicinity of Woodland Park, the Osburn Dam on the South Fork near Osburn, and the Pine Creek Dam was erected on the South Fork near the mouth of Pine Creek, not far from the Cataldo Mission. While these structures succeeded in holding back significant quantities of tailings for a number of years, subsequent floods, particularly those of December 1917, ruined the dams, causing great quantities of tailings to wash downstream. The MOA never rebuilt the dams. Millions of tons of tailings remained in place on floodplains above the now-defunct dams. Some of these tailings sit there to this day.

The flooding of 1917 and consequent destruction of the MOA's dams was at least partly responsible for the next series of environmental lawsuits to confront the mining interests. Over the next decade sixteen actions were filed against the mine owners in US District Court. Downstream rancher Jacob Polak filed the most legally significant of these. In 1919, Polak, owner of 320 acres near Cataldo, brought suit against the Hecla, Gold Hunter, and Bunker Hill mining companies. Polak charged that the long-standing movement of mining companies' tailings down the South Fork of the Coeur d'Alene had substantially filled the natural channel of that watercourse, in turn making the flooding of the South Fork a far more common event. This flooding, alleged Polak, brought poisonous, leaded waters to his land, crops, and livestock, damaging them all. Polak and other downstream farmers and ranchers also contended that the poisoned floodwaters had contaminated their groundwater and their domestic wells.[31]

In a rare legal defeat for the big mining companies, Judge Dietrich this time found Bunker Hill and its fellows responsible for property damages and awarded Polak $3,500 in compensation. The judge stated that despite its impoundment efforts, Bunker Hill's effluent into the South Fork contained fine materials. Dietrich also cited the defendants as negligent in their maintenance of Pine Creek Dam. In response to the mining companies' protestations that the construction of a series of settling ponds to contain their tailings was unfeasible, the federal judge responded that this both could and should be done, writing, "It follows that the problem is one of financial economy and not physical impossibility."[32] While granting the economic importance of the Coeur d'Alenes mining interests to the State of Idaho, Dietrich did not allow that this status conferred upon the companies the unlimited right to disparage the property rights of others. The judge wrote that the "mining industry is important to Idaho but the industry must bear its own burden, nor is it exempt from the fundamental maxim, '*sic utere tuo ut alienum non laedus*' ('use your own property in such a manner as to not injure that of another')."[33]

Upon appeal to the Ninth Circuit Court of Appeals in San Francisco, the court upheld the district court's decision, ruling that the Idaho Constitution's preference for mining did not grant mining interests "the right to inflict unlimited injury upon property of those who have acquired vested rights as manufacturers or agriculturists."[34] The court also found that the evidence presented "shows beyond question" that the mining companies had acted in concert in discharging waste material into the river: purchasing pollution easements, building tailings dams, and maintaining an organization (the MOA) in which they acted as a unit for the deposition of mining and milling waste.[35] The mining companies, unwilling to accept a negative judgment of such significance, then petitioned the US Supreme Court for a writ of certiorari. The writ was denied, however, preserving the judgment.

The landmark Polak case apparently demonstrated that the mining interests' days of doing whatever they wanted with the South Fork were at an end. Or did it? While the law now stood in the way of Bunker Hill and other companies' indiscriminate harming of the economic interests of those downstream, what if the mining companies simply bought out those interests? In an era when courts did not perceive any legal rights to inhere in unclaimed nature, when only damages to vested property rights in nature (e.g., water rights, land ownership by farmers) could find legal remedy, if the Mine Owners Association simply "bought the problem" by purchasing all of the lands adjacent to the creeks and rivers it fouled, would there then be any legal barrier to the continued pollution of the South Fork of the Coeur d'Alene and its tributaries?

This was precisely the question that courts would face in the coming years. *Hill* evidently convinced Bunker Hill and a number of its counterparts that the time had come to buy out those who might be in a position to challenge their pollution of the stream. The *Polak* decision seems only to have strengthened the mining companies' commitment to this modus operandi. Beginning in 1910 the MOA began a program to indemnify property owners along the South Fork and the main stem of the Coeur d'Alene River against possible damage resulting from the overflow of sediments from the mining district.[36] MOA member companies shared the costs of these purchases according to their share of the group's total tonnage milled, meaning that the most productive members, such as the Guggenheim-controlled Federal Mining and Smelting Company, the Day Mining Company, and Bunker Hill, paid out the most. Any land along the watercourse considered to be definitely threatened by the likelihood of overflow either was covered with an easement that "released the mines from all past and future pollution claims" or was purchased outright. The easements also applied to any and all "damage to crops, sickness, disease, or death to domestic animals" that upstream mining and milling operations might cause.[37] Using this stratagem, the mining interests gained either outright control or usufruct rights to pollute nearly all the land abutting the South Fork and the main stem of the Coeur d'Alene River from the mining district to Lake Coeur d'Alene. By the late 1930s roughly twenty-six thousand acres had been enrolled in this program.

From strictly a business standpoint, the mining interests' decision to buy their way out of the trouble posed by the threat of pollution lawsuits from downstream interests proved remarkably successful, at least until the political and legal shifts of the 1960s and 1970s that conferred protection and legal rights upon unowned nature altered the terrain of the game. Following the *Polak* decision in 1925, there were no successful lawsuits by downstream interests against the mining companies, though a number were attempted in the late 1920s and early 1930s. It appeared that the big mining and milling companies of the Coeur d'Alenes such as Federal, Hecla, Day Mines, and Bunker Hill—it has been estimated that sixteen of the area's more than one hundred mills contributed 93 percent of the total tailings to the streams—had solved their legal problem over water pollution without having been forced to alter their milling practices or substantively address the issue of the pollution of the Coeur d'Alene River watershed.[38] It should be noted, however, that in a small exception to the rule of mining company inaction the Bunker Hill Company, owner and operator of the district's largest concentrating operation, built the district's first tailings impoundment in 1928. Located between its Kellogg mills and the South Fork of the Coeur d'Alene, the impoundment was filled to capacity by 1956, when it was replaced by another one.[39] Though Bunker's tailings facility undoubtedly represented a step forward with regard to the pollution of the watershed, it could hold only a fraction of the district's tailings and still leached toxic heavy metals into the river system.

In the early 1930s a significant challenge to mining interests' treatment of the Coeur d'Alene River watershed arose from a different arena. In late 1929 and early 1930, *Coeur d'Alene Press* editor John Knox Coe fanned antipollution flames with a series of articles. The immediate trigger for the Coeur d'Alene City editor's muckraking series apparently was the discovery of sixteen dead muskrats in Thompson Lake, one of the numerous lateral lakes near the river's mouth at Harrison. With concerns mounting that the lake's water had been contaminated by upstream mining wastes, lake water was sent to Boise for testing to determine its arsenic levels.[40] Employing a vivid style, Coe attacked what he depicted as the upstream mining interests' wanton despoliation of the once-idyllic Coeur d'Alene River valley and the rising threat to the health and beauty of scenic Lake Coeur d'Alene. Four decades before the first Earth Day, Coe placed the ecological health of the Coeur d'Alene River watershed at the heart of a classically declensionist environmental narrative. The first article in the series appeared on *Press*'s front page on December 23, 1929. Titled "Once It Was a Vale of Plenty but Now—Coeur d'Alene River Valley Is Desolate; Party of Officials and Press Make Exploration Trip," the article depicted a recent boat trip taken by the Kootenai County commissioners, a state legislator from Kootenai County, and Coe himself, up the lake from Coeur d'Alene City to the inlet of the Coeur d'Alene River at Harrison and then up the Coeur d'Alene River.

The voyage's explicit purpose was to ascertain the extent of the mining-inflicted pollution of the lake and river. Coe quite literally ran across more evidence than he had hoped to find when his boat ran aground on a tailings bar sitting several hundreds of yards out into the lake. The editor wrote, "The hour spent in extricating the boat from the mess of mud, slime and mill refuse in the midst of a stench that was almost stifling . . . was the best evidence of what is actually taking place in the Coeur d'Alene River that could ever be found."[41]

Even before running aground, however, Coe and his party saw evidence of pollution. By the time they reached East Point, three miles downlake from Harrison, Coe detected "an appreciable coloring to the water. It assumed the appearance of yellow-grayish glass."[42] At Harrison passengers could see both the Coeur d' Alene River and the Saint Joe River, which hosted little mining activity, entering the lake. Coe contrasted the appearance of the two streams, writing, "There was a clear line of demarcation between the water coming down from the St. Joe River; clear, sparkling, and clean, and that emanating from the Coeur d'Alene; muddy, yellowish and revolting with a noticeable stench arising therefrom."[43] Coming to the lateral lakes that lie adjacent the Coeur d'Alene River, Coe worried that they, "with one or two exceptions," were being polluted by mine wastes due to "the slime infested waters pouring into the lakes."[44] Continuing upriver, Coe noted the decrepit state of many once-proud farms and ranches, writing, "Buildings that were once imposing homes, large hay barns and cattle barns telling of prosperous years, large incomes and happy community life, all are sad reminders of what once was. The buildings have collapsed, windows, doors and fences are all wrecks of their former selves."[45]

In Coe's next article, "Valley of Desolation Tells Its Own Story," the editor detailed heart-rending stories of individual farmers and ranchers in the "Valley of

Death," whose lands had been poisoned and who had been ruined by "the octopus of heartless wealth." Of this sad trend the editor wrote, "The list of former prosperous land owners who lost everything is almost endless. In varying degrees all suffered. A few with hill lands in conjunction managed to hang on. The great majority had to give up."[46]

For his final article in the series, "Once a Fruitful Paradise—Now the Valley of Death," Coe contrasted the recollections of old-timers such as Armand Perrenoud of a relatively pristine Coeur d'Alene River valley with the contemporary reality. The following month, Coe wrote a front-page coda to the series. In it, the editor suggested that all was not lost. The mining companies still could take effective measures to deal with the pollution of the Coeur d'Alene River and Lake Coeur d'Alene—either voluntarily or at the behest of the state legislature—the choice was theirs. The editor went on to suggest a technological remedy for the tailings problem: the construction of settling ponds below the various milling operations of the mining district. Coe indicated the possibility that the mining companies would take such steps voluntarily "if properly approached," but also suggested a willingness to take the issue to the political arena if necessary, writing, "And if they will not stop of their own free will it is possible to force them to stop their action in the state legislature."[47] For those legislators reluctant to enact novel pollution-control legislation, Coe cited as examples the stream pollution acts recently made law in Illinois and Wisconsin. Of course, 1930 was an election year. Coe, along with other Kootenai County residents, was poised to make a political issue of the mine wastes flowing from the upstream mining district.

Perhaps to avoid adding fuel to the fire, the mining companies remained notably quiet in the face of Coe's crusading series, undoubtedly buoyed by the long ascendance in the State of Idaho of the laissez-faire philosophy that government intrusion into the workings of business should be kept to a bare minimum. If there were pollution problems—and the mining companies were by no means willing to make any such admission—then they were to be dealt with by private industry on its own terms.

Kootenai County Republicans, however, had other ideas. Dependent primarily upon logging, farming, and tourism—not mining—for their local economy and tax revenues, they had little to gain from bowing to the concerns of Shoshone County mining interests. At their August 1930 convention, the Kootenai County Republican Party passed a resolution concerning "the constant dumping of mine tailings and slimes into the Coeur d'Alene River."[48] Echoing the tenor of Coe's series of articles, the Kootenai Republicans declared that the disposal of mine tailings had rendered the river valley barren and economically unproductive. This destruction had in turn, they stated, cost the county a great deal in lost tax revenue. The resolution further contended that the continuance of the mining district's methods of dealing with its wastes represented a "menace" to Lake Coeur d'Alene. The Kootenai GOP leaders directed county commissioner nominees, if elected, to appropriate funds for the purpose of taking the county's case to the appropriate judicial jurisdictions. Finally, the local Republican leaders directed candidates for state senator and representative to initiate legislation to protect Kootenai County citizens from mining pollution and to guard against "any further destruction of their property and rights."[49]

The Coeur d'Alene River and Lake Commission

The twenty-first session of the Idaho legislature convened on January 5, 1931. The Counties Committee took up the issue of water pollution in the Coeur d'Alene watershed, with Kootenai County senator Ralph S. Nelson leading the discussion. In the first week of the session, the state chemist and sanitary engineer for the Department of Public Health, W. V. Leonard, presented to the committee his findings from a November expedition to Lake Coeur d'Alene and the Coeur d'Alene River. Leonard's report suggested that while no lead or lead salts in solution were found in the lake, suspended tailings were carried in the river. The report also highlighted a significant quantity of bacterial pollution issuing from the disposal of raw sewage into the river.[50] Admitting that his study was not conclusive due to time constraints and the seasonally low flow of the river when he observed it, the state chemist recommended that a commission be established to further study the situation.

On March 16, the Idaho legislature approved an emergency act creating the Coeur d'Alene River and Lake Commission. The commission consisted of the state attorney general, Fred J. Babcock, who acted as chair, along with the chairmen of the boards of county commissioners from Kootenai and Shoshone Counties, E. O. Cathcart and James H. Taylor, respectively. The act's emergency status permitted the commission to begin using state funds immediately. The legislature granted $3,000 for the publishing of a report and expenses. The report was to be delivered to the next session of the legislature, at the beginning of 1933. The act required the commission to investigate the "ways and means of eliminating from the Coeur d'Alene River and Coeur d'Alene Lake, so far as practicable, all industrial wastes which pollute[d]" them, and to recommend means of preventing further pollution that would be harmful to vegetation, domestic crops, public health, or animal life, including fish and aquatic life.[51] It offered the services of W. V. Leonard and provided for the opportunity to employ the services of other technical advisors as needed.

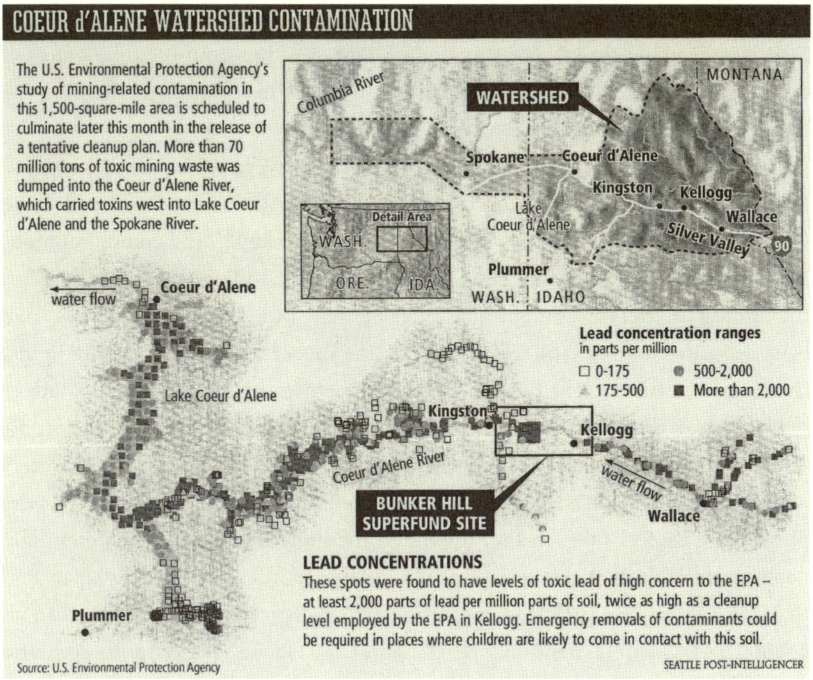

FIGURE 3.2. Coeur d'Alene Watershed, Lead Contamination (US Environmental Protection Agency).

At its initial meeting in May, the commission elected Attorney General Babcock its chair and opted to contact the US Biological Survey and the Federal Fish and Game Commission to request the assistance of a federal sanitary engineer to work with Leonard in studying the condition of the Coeur d'Alene River and Lake Coeur d'Alene. The commission held its next public meeting at Rose Lake, near Harrison, on August 29. About sixty people showed up, among them a number of Kootenai County farmers who testified about the death of horses and cattle that had grazed upon meadowlands that had been flooded by river waters.

The commission eventually enlisted the aid of the US Public Health Service, the US Bureau of Fisheries, the federal Bureau of Mines, and the Idaho Department of Public Welfare. All of the involved agencies except the Bureau of Fisheries coordinated their efforts so as to avoid duplication. The Department of Public Welfare and the Public Health Service, under the direction of J. K. Hoskins, sanitary engineer in charge of the stream pollution station at Cincinnati, Ohio, investigated the pollution of Lake Coeur d'Alene. The Bureau of Mines studied the source pollution—the mining district—along with the length of the South Fork, and the main stem of the Coeur d'Alene River. In a separate investigation, the Bureau of Fisheries' head of fisheries investigations, Dr. M. M. Ellis, examined the waters of the Coeur d'Alene River system from above Wallace to the Spokane River at the Idaho-Washington state line.

The commission's report, submitted on December 24, 1932, constituted a potent, science-based statement on the state of the Coeur d'Alene River and Lake Coeur d'Alene following nearly a half century of mining.[52] The commission observed that in the period since the flotation system had been widely adopted in the mining district, "enormous quantities of finely powdered rock and ores have been deposited in the channel of Coeur d'Alene River and on lands adjacent to this river in the Coeur d'Alene River valley, from the town of Cataldo to the mouth of the river near Harrison, where it empties into Coeur d'Alene Lake."[53] The report also found that lead was present, both in suspension and in solution, in the waters of Lake Coeur d'Alene and Coeur d'Alene River.[54] Of the two, however, it was the presence of lead in solution that presented the more intractable and potentially dangerous problem. While it was readily feasible to filter out lead in suspension, that was not the case with lead in solution, which also was far more toxic than its counterpart. Coeur d'Alene City relied upon Lake Coeur d'Alene for its domestic water supply and the town of Harrison relied upon the Coeur d'Alene River.

The commission noted with apparent surprise its discovery that the lead content of river and lake deposits tended to be higher than those of mill tailings found in the mining district, and concluded that this likely resulted from the repeated rewashing, during periodic flooding, of old downstream tailings that contained a higher percentage of lead than did current tailings.[55] Turning to the problem of lead in solution, the commission accepted Ellis's judgment that its presence was due to the continued exposure of old streamside tailings deposits to weathering processes. When flooding occurred, as it often did, and these deposits were rewashed, lead in solution wound up in the river and lake.[56] Describing this recurrent problem, the report noted, "The Coeur d'Alene River periodically overflows its banks and creates a settling basin of some 25,000 acres upon which this lead or other mineral may be settled and is probably the chief contributing factor to the problem of lead in solution through the process of weathering."[57] By 1932, roughly twenty-five thousand acres of land lying adjacent to the South Fork and the main stem of the Coeur d'Alene River had, in the parlance of local farmers and ranchers, become "leaded." In an increasingly toxic cycle, with each new winter or spring flood already-polluted lands added additional heavy metals to the downstream environment. It was clear that, by the early 1930s, if not earlier, the Coeur d'Alene valley and lake region found itself caught in a spiraling cycle of environmental pollution.

Hoskins and his team took numerous water samples from various locations in Lake Coeur d'Alene, and from one location in the lower Coeur d'Alene River and found that "under normal conditions the lake water is practically saturated continuously with lead in solution."[58] Hoskins discovered that while most of the water samples' lead contents ranged from 0.08 to 0.20 parts per million (ppm), a few indicated the presence of lead in much higher quantities. A sample taken at the Harrison intake contained lead at a ratio of 2.25 ppm and one at the Coeur d'Alene City intake showed a lead level of 1.75 ppm. In the early 1930s federal drinking water standards (at the time maintained by the US Department of the Treasury) specified that lead should not exceed 0.1 ppm. Most Lake Coeur d'Alene water exceeded the federal standard.[59] Hoskins found that the lake water's lead levels tended to decrease with distance from the mouth of the Coeur d'Alene River, with lower levels found at Coeur d'Alene City than at Harrison. The sanitary engineer attributed this to the settling out of lead to the lake's bed as water traveled from the mouth of the river to the point near Coeur d'Alene City where the lake emptied into the Spokane River.

Based upon the average lead reading of 0.22 ppm at the Coeur d'Alene City drinking water intake and the estimate that it would take the absorption of 1/32 of a grain of lead daily to "set up chronic lead poisoning,"[60] Hoskins calculated that a person would have to drink 2.4 gallons of water a day—far more than the average person consumed—to be in danger of acquiring chronic lead poisoning. He did not perform such calculations for the city of Harrison, where the average lead content of the water at the intake was found to be 0.36 ppm, more than 50 percent higher than at Coeur d'Alene City. Had Hoskins done so, however, he presumably would have concluded that a Harrison resident could safely drink only about 1.6 gallons of water per day, a far less comforting amount, especially if one factors in the amount of water used daily in cooking. Employing early twenty-first-century standards, under which they would have been considered at risk of contracting lead poisoning by drinking as little as one-fifth of a gallon of water per day at Coeur d'Alene City and one-eighth of a gallon at Harrison, residents of these communities likely would not have felt safe in their consumption of local drinking water.

The Bureau of Mines, whose fundamental mission was to act as a governmental handmaiden to the mining industry by providing it with useful advice and scientific data, was, of the federal agencies involved in the investigation, the one most accurately perceived as "a friend of the mining industry." It was not surprising, then, that the bureau's report stood out for its minimization of the "alleged" harm done to local flora and fauna by mine tailings and did the most to buttress the mining industry's "technological fix" to the water pollution problem—the massive suction dredge that the MOA installed at a spot near Cataldo where the valley widens and the river's pace slackens. The mine owners set up the dredge in 1932 while the investigation into the pollution of the Coeur d'Alene River and lake was ongoing. Rejecting the option of establishing settling basins at Cataldo and piping tailings there directly from the concentrating mills as overly expensive and technologically unfeasible, the mine owners instead decided to allow their tailings to continue to trail down the South Fork and the main stem of the Coeur d'Alene River to Cataldo before removing and stacking them on adjacent land with an electric suction dredge. Commenting upon this proposed solution, the bureau opined that "there are no apparent reasons for believing the dredge will not be an amelioration, and ultimate improvement can be expected. Plants that operate on the same principles for removing tailings from water have proven to be satisfactory in other situations." The bureau did not mention the fact that the dredge would do nothing to ameliorate the condition of the South Fork or the brief stretch of the Coeur d'Alene River above the dredge at Cataldo.

The most noteworthy and ecologically based of the federal reports was Ellis's. In his sixty-one-page report, the Bureau of Fisheries scientist assessed the quality of the South Fork, the Coeur d'Alene River proper, and Lake Coeur d'Alene as habitat for flora and fauna and made recommendations for improving their respective conditions. Based upon his studies of the watershed in July 1932, Ellis found the South Fork, from a few miles above Wallace, and the entirety of the main stem of the Coeur d'Alene River, to be essentially a dead river, describing it as "practically devoid of fish fauna, bottom fauna or plankton organisms."[61] Without plankton, fish life was practically an impossibility, so it was hardly encouraging to learn that the roughly fifty miles of river from above Wallace to Harrison lacked plankton life. Further discouraging was Ellis's discovery that the plankton fauna of Lake Coeur d'Alene was "as a whole rather sparse, and particularly poor at the south end."[62] Ellis found that while the lake's plankton conformed to essentially the same distribution patterns observed in fisheries biologist George Kemmerer's studies of 1911, they had declined during the intervening twenty-one years of mining discharges.[63]

As part of the US Bureau of Fisheries' study of the fish production potential of the Northwest's lakes, Kemmerer had examined the biological and chemical properties of Lake Coeur d'Alene. He noted that the Coeur d'Alene River's waters as they entered the lake at Harrison "are so laden with silt that they may be traced far out into the clear water of the lake, the bottom of which shows the effect of the sediment from both rivers [the Coeur d'Alene and the Saint Joe]."[64] Kemmerer also had seen evidence that the Coeur d'Alene River was dropping much of its mineral-heavy load near the river's outlet into the lake, thus considerably decreasing the lake's depth there.[65] Kemmerer and his colleagues noted the thinness of plankton at the river's opaque outlet.[66] His catches in the deepest part of the lake also indicated a relative lack of plankton.[67]

The intervening twenty years had hardly improved the lake's health. Ellis found negative trend lines in Lake Coeur d'Alene's bottom fauna, describing it as "very meager" and noting that it "seemed to be declining."[68] He found no living fish at the mouth of the Coeur d'Alene River, but detected the presence of minnows, dace, bass, perch, and trout in other parts of the lake. Ellis noted, however, that trout were rarely caught off Harrison at that point, and that their numbers across the lake seemed to have decreased significantly since the time when Captain Mullan visited the lake in 1858 and found it to be "a noble sheet of water . . . filled with an abundance of delicious salmon trout."[69]

Addressing the scientific reasons behind the absence of aquatic life, Ellis first focused on the state of the bed of the Coeur d'Alene River. He found a riverbed that "was deeply covered with shifting deposits of very fine rock powder which not only overwhelmed the bottom fauna but prevented the development of any bottom consocies of animals either adult or immature."[70] Effective killers, the vast quantities of mine tailings that had washed downstream not only had choked out existing life in the river but also had prevented the development of new life. Examining the deposits of mine slimes on lands adjacent to the Coeur d'Alene River about which litigating farmers had often complained, Ellis wrote, "Along the polluted portion of the Coeur d'Alene River the banks, flats, and low lands were covered often to a depth of several feet with deposits of mine slimes which had been left there during high water and which were being continually returned to the river by wind, rain, and current action, *thus producing a constant repollution of the river* [italics added]."[71] Here Ellis found highly toxic "incrustations" that often formed on the mine slimes and on soils that were "more or less impregnated with mine slimes"[72]—hard coatings over the soil that often covered several acres. Determining these incrustations to be "composed of a soluble fraction [3 to 40 percent] which was largely zinc sulphate and an insoluble fraction carrying zinc, and lead compounds," Ellis and his team performed experiments on animals. The experiments showed that the soluble fraction of the incrustations "was highly toxic to fish, frogs, turtles, fresh-water mussels and plankton crustacea."[73] After exposing fish to high-concentration dilutions of these incrustations for twelve to thirty-one days, the fish died, "showing among other symptoms black deposits of lead in the fins and elsewhere about the body, indicative of cumulative lead poisoning."[74] Rewashed mine tailings, it turned out, were highly lethal to fish—bad news for "the finny tribe" in the Coeur d'Alene River, a water body that remained prone to flooding.

Conducting further experiments, Ellis placed long-nosed dace and dace minnow caught in Lake Coeur d'Alene in live buckets at two separate locations, one in Lake Coeur d'Alene offshore Harrison Beach, the other about one-third of a mile up the Coeur d'Alene River from its mouth at Harrison. After twenty-four hours both buckets were raised and all of the fish seemed to be in fine condition. The buckets then were placed back in the water. After seventy-two hours the buckets again were raised. This time while all of the fish in the Harrison Beach bucket "were active and in good condition," all of the fish in the Coeur d'Alene River bucket were dead. In addition, each of the dead fish "was covered with a heavy coating of mucous slime and the gills were choked with mucous."[75]

Ellis concluded that the death of the Coeur d'Alene River fish could not be ascribed to "low oxygen, high carbon dioxide, high osmotic tension or a temperature differential . . . because there were no deaths in the first 24 hours." To the fisheries scientist the "heavy secretion of mucous" by the fish clearly pointed to an irritation, "either chemical or mechanical, or possibly both, since the secretion of mucous is one of the protective reactions of living fish."[76] At the same time, the fish placed in the lake, approximately a third of a mile away from the other group, apparently were doing fine. The chief difference in the situations for the two groups of fish seemed to be the relative concentration of mine wastes. In the lake the wastes were given more of an opportunity to diffuse, whereas in the river, even thirty-five miles downstream from the mining district, they remained relatively concentrated.

Turning to the effects of area waters on local plankton, Ellis conducted experiments using water from the Coeur d'Alene River near Dudley and from Lake Coeur d'Alene near the mouth of the river. As a control group, Ellis placed some plankton in water from the headwaters of the South Fork, where healthy populations of plankton were found to exist. All of the plankton exposed to river water died in eighteen hours or less, while the plankton exposed to Lake Coeur d'Alene water fared slightly better—most survived for forty-eight to seventy-two hours before succumbing, and about 20 percent survived outright for three days.[77] These results largely conformed to the pattern found in the river and the lake, where there was no plankton presence in the Coeur d'Alene River and a sparse presence at Harrison.

To ensure that his experiments tested only the harm caused by substances present in solution in the Coeur d'Alene River and Lake Coeur d'Alene, Ellis removed the material present in suspension before testing the waters. Thus he could rule out the "mechanical action" (i.e., abrasion or crushing) of the waters as a cause of harm to plankton. Ellis concluded "that the Coeur d'Alene River water contained salts or other substances in solution during July 1932, which were quite toxic to plankton animals."[78]

Ellis devoted considerable attention to the "lead incrustations" that formed such a prominent landscape feature on the lands directly adjacent to the Coeur d'Alene River from Cataldo down to Harrison. His interest was due principally to the fact that, as substantial portions of them consisted of water-soluble lead and zinc sulfates, as opposed to insoluble sulfides, the incrustations had the potential to be highly toxic to plant and animal life and so could be central to the toxicity of the Coeur d'Alene River. The incrustations also might help explain the perplexing reality that the river water contained higher levels of lead below Cataldo, fifteen miles downstream from the heart of the mining district, than it did directly below the district. Ellis first determined that the incrustations had been produced by the weathering of exposed mine wastes, "which are subjected to a variety of conditions conducive to chemical changes [such] as moisture, light, heat, and oxidation." In appearance, the biologist described the incrustations as "of a crystalline material varying from pure white to dull gray or dirty brown in color." Their extent was staggering. Ellis spoke of the gross bulk of the incrustations as amounting to "the enormous weight of 27 to 28 tons per acre of surface incrusted." Ellis noted that particularly far-ranging examples of incrustations were observed at "flats" near Dudley and Thompson Flat and also that "practically every exposed mass of mine slime deposits along the river from Cataldo to the mouth, was more or less covered with these incrustations."[79]

Performing chemical tests upon the incrustations at Thompson Flat, Ellis found that 3.3 percent of the incrustations, by total weight, were water soluble and that this soluble fraction was composed of 62 percent zinc sulfate, 30 percent other soluble solutions—principally sodium and magnesium—and 0.05 percent lead. The fisheries biologist also conducted experiments to determine what effects this water-soluble fraction would have on the relatively hardy common goldfish. When the water-soluble solution was present in the water with goldfish at a level of 0.01 percent or stronger, "a profuse secretion of mucous by the fish was induced in the course of the first 6 hours or less." This was a sign of bodily irritation in the fish. At such a low level of exposure, the goldfish stopped secreting mucus after four days and appeared perfectly healthy again. At higher levels of exposure, though, the goldfish evinced greater physical difficulties including the more voluminous production of mucus, sluggishness of movement, and difficulty swallowing food. When the solution was introduced on a daily basis, as might occur on the Coeur d'Alene River during a flood, all of the fish perished within four to six days.[80]

Despite the relatively low level of lead present in the water-soluble fraction of the incrustations, Ellis found it to be the culprit in the deaths of the goldfish used in his experiments. While zinc, present in much higher levels in the incrustations, is not a cumulative toxin, lead is. Ellis determined that the zinc present in the incrustations irritated the fish, causing mucus production, but did not kill them.[81]

Ellis also tested the reactions of goldfish to lead sulfide.[82] Although minute amounts of lead precipitated out over time when the sulfide was placed in water, this substance was many orders of magnitude less water soluble than lead sulfate. This distinction mattered both legally and ecologically. In *Doty* the presiding judge accepted the mining companies' claims that the lead ore they produced, lead sulfide, was harmless, based upon, among other things, the companies' claim that cows that drank the water directly from their mill flumes had grown fat and happy. *Doty* released the companies from any liability for the crop and livestock losses of downstream farmers. Ellis's experiments revealed that, indeed, goldfish had little to worry about from lead sulfides in the form of washed lead ore. While nearly all of the goldfish continually exposed to lead incrustation material, lead sulfate, had perished within thirty-one days, the goldfish exposed to washed lead ore, lead sulfide, all survived, apparently quite healthy, for the same period of time.[83] Ecologically, the problem for the Coeur d'Alene watershed lay primarily with lead sulfate—lead compounds following their exposure to weathering action and return to the river—and not with lead ore per se. A significant portion of lead ore tailings, however, after flowing downriver from the mining district, invariably washed onto riverbanks and flats and began the process of weathering that converted them into highly toxic lead sulfates. Unless these lead tailings were somehow collected and stored in an impervious container, or were captured directly below the mills and placed in impoundment ponds, it was unclear how the huge ecological problems they were causing could be solved.

To determine their toxicity for other fauna that frequented the Coeur d'Alene watershed, Ellis also tested the effects of the incrustation material on frogs, turtles, and freshwater mollusks. He found that, even highly diluted, the material had a dramatic, negative effect on the health of the creatures. It "produced immediate paralysis of the stomach and intestines of fishes, frogs and turtles, in dilutions as high as 1:100,000." This paralysis unfortunately was "very largely irreversible." The material also "caused immediate disturbances in the heart beat of fishes, frogs, turtles and fresh-water mussels which shortly resulted in complete cessation of heart beat, when dilutions as high as 1:100,000 were used."[84] Ellis concluded that the incrustations were "highly toxic to . . . frogs, turtles and fresh-water mussels if the exposure were long enough to allow even very small quantities of this soluble fraction to be absorbed, taken into the alimentary canal, or otherwise brought to vital organs."[85] Little hope existed for the return of land animals that lived on or by the banks of the South Fork and the main stem of the Coeur d'Alene River as long as significant amounts of the incrustation material remained.

Moving downstream to Lake Coeur d'Alene, the large settling basin for the waters of the Coeur d'Alene and Saint Joe Rivers, Ellis found much evidence of the effects of the mining district's wastes. Dredging material from the bottom of the lake at numerous points revealed that mine slimes had nearly blanketed its floor. The presence of these slimes, Ellis concluded, was a significant factor in the relative sparsity of bottom creatures living on the lake floor. This lack of bottom life was significant since such life constitutes a cornerstone in aquamarine food chains. The fisheries biologist also noted the presence of mine slimes along the shores of Lake Coeur d'Alene and even up relatively high on a prominent shore feature known as Black Rock, demonstrating that the fine tailings produced by the flotation process were making their way all around the lake.[86] Ellis even found evidence of mine slimes as far downstream as Green Acres, Washington, along the Spokane River, a number of miles below the lake.[87]

At the end of his report Ellis summed up his findings and made recommendations for the restoration of the fish populations of the Coeur d'Alene River and Lake Coeur d'Alene. He observed that "collectively the mine wastes in the Coeur d'Alene River have destroyed the fish fauna and the plants and animals on which fishes feed," and that "these conditions in the Coeur d'Alene River not only have made that river barren, but they are threatening all of the lakes adjacent to the Coeur d'Alene River, and constitute a constant menace to Coeur d'Alene Lake." Based upon his studies, the Department of the Interior's chief of fisheries' investigations concluded, "There is only one complete solution to this problem and that is the exclusion of all mine wastes from the Coeur d'Alene River." Even with such drastic curtailment of pollution, however, Ellis estimated that "it will take the river some time to rid itself of the masses of powdered rock in the stream bed and of the weathering deposits of mine slimes along its banks, so that several years will elapse before a natural fauna will reestablish itself in the now polluted portion of the Coeur d'Alene River."[88]

Somewhat more realistically, given the political economy of the Coeur d'Alenes, Ellis also proffered a technological solution that did not involve the complete shutdown of the area's mines and mills. Based upon his recent exploratory visit to the comparably sized lead and silver Sullivan mines and mills at Kimberley, British Columbia, the fisheries biologist recommended that the Idaho mining companies adopt something akin to the filtration system in place in Kimberley.

That system employed wooden flumes to transport mine wastes to a number of settling basins where mill tailings progressively settled out. The settling basins also progressively filtered the mill water so that, per Ellis, by the time it reentered the Saint Marys River, "the mine chemicals and other substances in solution are very largely destroyed" and "plankton, algae, aquatic insects and fish were found."[89] Ellis also noted that the Kimberley filtration system—essentially mandated by Canadian water laws that prohibited the dumping of mill wastes into the nation's natural waters—had not been cost-prohibitive to construct and involved only nominal costs to maintain. Turning to the suction dredge, the likely rival pollution-control system that the MOA recently had put into operation on the Coeur d'Alene River, Ellis granted that it promised an improvement in the water quality of the Coeur d'Alene River and lake, but concluded that it would fall far short of restoring the health of those waters, stating, "This procedure will undoubtedly reduce the amount of mine slimes carried to the lower river but as the finer particles will not settle out even in Coeur d'Alene Lake the pollution of the lower river will not be corrected but merely modified."[90] Short of shuttering the Coeur d'Alenes mines and mills, Ellis clearly saw the installation of something akin to the Kimberley settling ponds system as the only reasonable hope of restoring the ecological health of the Coeur d'Alene watershed.

As it happened, Ellis's technological recommendation became the principal sticking point for the Coeur d'Alene River and Lake Commission. In their final report, two of the three commissioners—Attorney General Babcock and Kootenai County Commissioner Cathcart—voted to approve the full report, including Ellis's conclusions in the "Recommendations" section. The third commissioner, however, Shoshone County Commissioner James H. Taylor, voted against approving the full report, noting his specific objection to the report's endorsement of the proposal to transport mine wastes to settling beds by some means other than the South Fork of the Coeur d'Alene River. Taylor opposed the use of settling ponds and instead supported the use of the dredge at Cataldo. It was "his position that the dredge as set up and operated at the present time will satisfactorily handle the question of slimes in the Coeur d'Alene River."[91] That Taylor's position mirrored that of the mining interests who played such a powerful economic and social role in the county where he practiced politics is not surprising. It is interesting, however, that Taylor did not publicly object to any other aspects of the report. Perhaps he felt unequipped to challenge the observations and conclusions of scientists working on their own turf but was more secure in tackling the technical engineering-based proposals of Ellis, a nonengineer. In any event, the Shoshone County commissioner's refusal to sign on to the final report meant that it lacked consensus and also that the upstream-downstream conflicts that spawned the commission were still in evidence at the conclusion of its efforts.

The River and Lake Commission, in fulfilling its duty to recommend means of preventing pollution that would be harmful to animals, plants, and human health in and along the Coeur d'Alene River and lake, made only two recommendations to the Idaho legislature. The first made favorable mention of the recently constructed dredge at Cataldo and suggested that the MOA's efforts should be applauded and use of the suction dredge continued, stating that "the use of the dredge may be very effective in cleaning the deposits of slimes out of the Coeur d'Alene River and reclaiming that portion of the river below Cataldo." This recommendation would have been rendered pointless, however, had its second recommendation—for installation of a flume and settling ponds system—been adopted. Ellis reported that the settling ponds system he observed in Kimberley had filtered the mine wastes so well that the water leaving the final pond was clean enough to sustain substantial aquatic life. It appears, then, that the commission essentially decided to hedge its bets in its recommendations to the legislature. Understanding that it was unlikely that the MOA would build a settling ponds system like that found in Kimberley after having already gone to the expense of purchasing and installing a large dredge at Cataldo, the commission bowed to reality and recommended what was already a fait accompli, noting that the settling ponds were optimal but that use of a dredge was still a good choice.[92] In the depths of the Great Depression, little opposition could be expected to mass against the recommendation. An ecological half measure, this nonetheless probably represented the best that could be achieved at the time given the opposition of north Idaho's powerful mining industry to further action.

The Twenty-Second Idaho Legislature accepted the commission's report in January 1933 and then proceeded to let it collect dust. The State of Idaho would take no further action regarding the pollution of the Coeur d'Alene basin for decades.

Federal Efforts and the Coeur d'Alene Watershed

In March 1933, in the midst of the Great Depression, a new administration took the reins of power in Washington, promising a "New Deal" for Americans. Franklin D. Roosevelt and his New Dealers swept into office and began an aggressive, often highly improvisatory effort to relieve the country's deep social and economic burdens. Roosevelt's more activist approach to governing led to the creation of the numerous "alphabet soup" agencies designed to give work to the unemployed and tackle a wide range of social and economic ills. One program of the Works Progress

Administration (WPA) was the Federal Writers' Project (FWP), which put quite a few underemployed writers to work on various projects, including a series of state guides.

Vardis Fisher, a native Idahoan and prolific novelist and essayist, wrote the FWP book on Idaho, *Idaho: A Guide in Word and Picture*, published in 1937. Noting the Coeur d'Alenes' productive scope and capacity, Fisher recorded that the area was the richest silver producer in the nation's foremost silver-producing state and that it possessed the largest silver mine (the Sunshine) and the three largest lead mines in the country.[93] However, the author also was unsparing in his portrait of the environmental despoliation he witnessed on his mid-1930s travels through the region. Part history, part travelogue, *Idaho: A Guide in Word and Picture* depicts Fisher's journey through the mining district on US Highway 10 (now Interstate 90). He crosses from Montana into Idaho at Lookout Pass and begins the downhill trek to Mullan, six miles distant, mentioning that the vistas found there "offer little foretaste of what is to be found in the Coeur d'Alene mining district." At Mullan, Fisher notes the inescapable presence of the Morning Mine, "the third largest lead producer in the United States," along with the fact that it "sustains most of the population of the town" of 1,891. Then Fisher turns his eye to the South Fork of the Coeur d'Alene, observing, "On the right, too, as the highway leaves the W[est] end of town, is the river, but it is not the lucid stream of a mile ago. It has been diverted to the mines here, impregnated with poison, and turned free. It now looks like a river of lye. Or, better, it looks as if all the dirty clothes in the world had been washed in it."[94]

By Fisher's account, the South Fork already looked badly tarnished at Mullan, even before it had traveled through the bulk of the mining district. Eighteen miles farther downstream at Kellogg, by this time the largest town in the district (pop. 4,124) and home to the giant Bunker Hill mine and smelter, Fisher's description of the state of the river turns more grim. He writes, "Below [the Bunker Hill mine] the river bottoms look like a caricature of a graveyard."[95] Traveling a bit farther down the South Fork, the author observes that "west of Kellogg with its miracles of machinery, there is still to be seen a poisoned or dead and dying landscape. Trees slain by the invisible giant still stand with lifeless limbs and with roots still sucking the poisoned earth."[96] Fisher's portrait of the South Fork of the Coeur d'Alene where it passed through the Coeur d'Alene mining district in the 1930s renders a watercourse that has been laid to waste.

Roosevelt and his New Dealers, however, wanted to do more than simply depict the nation's water pollution, much of which had been caused by mining operations and a lack of adequate sewage treatment facilities. In 1934, Roosevelt used an executive order to create a National Resources Committee designed to study pollution, among other subjects. Its Advisory Committee on Water Pollution collected information from the states and various federal agencies. The committee published a report, *Pollution in the United States,* that made recommendations for pollution legislation. In 1939, Congress passed a major water pollution act that would have created a permanent national water resources committee to study pollution and make recommendations to the federal government for its abatement. Roosevelt reluctantly vetoed the legislation, however, because it lacked a permanent system for disbursing federal grants to states. A similar water pollution bill narrowly missed passage in 1940 and then World War II intervened, putting pollution concerns on the back burner. But in 1948, national politicos returned to the issue. In that year Congress passed, and President Harry S. Truman signed, the landmark Water Pollution Control Act of 1948, the nation's first water-pollution-control legislation.

The act provided for coordinated comprehensive planning, technical services, and research to address the states' water pollution problems along with their water resource potentials, under the auspices of the Public Health Service (PHS) and the surgeon general.[97] Although it was weakly funded in its initial years of operation, federal and state officials nonetheless went to work to fulfill the act's mandate to study the nation's water resources. For purposes of effectiveness and efficiency, the PHS divided the nation's waters into fourteen basins, for the most part corresponding to major watersheds such as the Columbia River. By 1951, the PHS, assisted by the Army Corps of Engineers and state and local officials, had created fourteen basin commissions and tasked them with reporting to Congress on a variety of issues relating to the nation's water quality and quantity.

The waters significantly affected by mining activities in the Coeur d'Alenes—the South Fork, the main stem of the Coeur d'Alene River, Lake Coeur d'Alene, and the Spokane River—all part of the sprawling Columbia River drainage system, were included in basin fourteen. This was the province of the Pacific Northwest Basin Commission, headquartered in Portland, Oregon. Unsurprisingly, the basin commission identified the mining and concentrating operations in the Coeur d'Alenes as one of the most significant threats to water quality in the Upper Columbia River basin, and the Coeur d'Alene watershed as one of the most heavily polluted in the region. Along with a handful of mining and concentrating operations in the upper basin, the commission identified the basin's rapidly growing population, and its scarcity of adequate sewage treatment facilities, as the region's chief water-quality threats.[98] Noting the environmental destructiveness of major mining operations such as those found in the Coeur d'Alenes, the Commission opined that "fortunately, the big mining operations—copper, lead, zinc, silver, gold, etc.,—have been concentrated in three relatively confined areas. In some localities these mining companies have worked out their own methods of control which have reduced the amount of wastes entering the watercourses. In other localities, however, such as . . . the south fork of the Coeur d'Alene River below Mullan, Idaho . . . large mining operations for years have degraded streams by the discharge of wastes directly to the watercourses."[99]

Despite what the Commission labeled as the mining companies' "excellent efforts" to mitigate their pollution of watercourses through technological solutions such as the establishment of dredging operations and the construction of tailings and settling ponds, along with their attempts to purchase the problem by buying land and pollution easements from downstream landowners, the commission concluded that "some 100 miles of watercourses have been destroyed for most normal uses."[100] Specifically addressing the effects of this pollution on municipal drinking water and aquatic life in the Coeur d'Alene watershed, the commission stated, "Wastes from ore concentration mills entering some of these watercourses have affected the drinking water supply of Harrison, Idaho; destroyed all fish and aquatic life in the South Fork of the Coeur d'Alene River below Mullan, Idaho and the Coeur d'Alene River below the mouth of the South Fork . . . In addition, the possibility of mine tailing wastes affecting the water supply of Coeur d'Alene, Idaho must be recognized."[101]

Echoing Ellis's conclusions of a decade earlier, the commission specified that "the inorganic wastes discharged to the streams are principally mine tailings, acids, and toxic chemicals. Probably the greatest stream degradations found in the area are caused by the discharge of tailings from ore concentration mills and acid ore mine wastes. These wastes contain many chemicals which are detrimental to normal water uses and cause depositions which destroy the aquatic life in streams." The basin commission, however, balanced the mining district's status as a grade-A water polluter with its importance to the nation's Cold War industrial and military strength, proudly noting that the Coeur d'Alenes constituted one of two mining districts in the basin (the other was the Butte-Anaconda, Montana, copper center) that had produced over a billion dollars' worth of mineral wealth, "a record achieved by only six other mining districts in the world."[102]

There was, however, no getting around the messy fact that mining—and even more so, concentrating—operations, centered on Kellogg and Butte/Anaconda, were seriously fouling the basin's waters and the lands adjacent to them. The commission noted that "ore concentration mills . . . are dumping annually more than a million tons of mine tailings into watercourses. Acid mine wastes are disposed of in the same manner." While conceding that, "pending development of practical disposal methods," it might be necessary to permit mine companies "to continue to use stretches of certain streams for receiving these wastes,"[103] the basin commission nonetheless maintained that in the future, "mine owners should be required to develop and provide the necessary treatment of their wastes."[104]

Shit Creek

The Columbia River Basin Commission Report of 1951 cited the lack of sewage treatment facilities to serve the roughly twenty thousand people who lived along the South Fork of the Coeur d'Alene River as another principal threat to the watershed. By 1960 nothing had been done about the situation and a group of concerned Lake Coeur d'Alene residents decided that it was time to address what they saw as a serious threat to the health of their lake and themselves. Forming the Coeur d'Alene Lake Anti-Pollution Committee, the group pressured Idaho governmental agencies and the Idaho Wildlife Federation to present recommendations to the Idaho legislature in 1961. The committee sought measures to prevent the continued dumping of raw sewage and mine wastes into water that eventually made its way down to the lake and also advocated for better enforcement of water quality regulations already on the books.[105] In the early and mid-1960s the upstream mining companies and local communities began making efforts to comply with federal and state clean water statutes. The valley's narrow, rugged terrain posed significant challenges, but by 1968, with Bunker Hill leading the way, the companies had constructed modern tailings impoundments and settling ponds at several locations in the valley to serve all the district's remaining mines.[106] The MOA decommissioned its dredge at Cataldo Flats. Responding to pressure from federal and state governments that were shedding their complaisance about polluted waterways, mining companies of the Coeur d'Alenes finally were complying with M. M. Ellis's recommendations from 1932.

Advocates of the installation of sewer systems and sewage treatment facilities in the Coeur d'Alenes, however, were having less success. In 1967 and again in 1968 local bond elections to finance the construction of sewage disposal infrastructure were defeated.[107] In 1968 the Idaho Department of Public Health began posting signs along the South Fork warning of sewage contamination.[108] When these signs began disappearing, the department responded by posting billboards threatening legal action. In 1969, the department persuaded area mining companies to join the regional sewer district and bear 60 percent of its costs. With their financial burden thus eased, local voters passed the sewage bond.[109] By 1972 almost no raw sewage was being funneled into the South Fork.[110]

Another technological improvement that benefited the Coeur d'Alene watershed was the adoption of the sandfill system. In 1949, the Dayrock Mine began extracting the sand fraction of its mill tailings and putting it back into the mine as fill material. Between 1954 and 1961 most other district mines adopted the technique. Mines generally used about half of their former tailings as fill. Given a steady rate of production, this meant the deposition of only about half as many tailings.[111] The adoption of the sandfill system, by necessitating the employment of less land in the narrow valley for the deposit of tailings, also freed up future space for the construction of settling ponds.[112]

A River of Many Colors

University of Idaho biologist Fred W. Rabe began studying the Coeur d'Alene watershed in 1967. Over the next seven years, with the help of his graduate students, Rabe extensively examined Lake Coeur d'Alene, along with the South Fork, the North Fork, and the main stem of the Coeur d'Alene River. Among Rabe's significant findings was that while the installation of tailings impoundments and settling ponds in the mining district undoubtedly made the South Fork and the main stem of the Coeur d'Alene River flow clearer, these measures did not necessarily make it cleaner. Fine clay particles that once had combined with heavy metal ions from the mining district to make them relatively inert now largely remained in the tailings ponds. Lacking, after 1968, these particles to bond with, the heavy metal ions that still found their way downstream were significantly more toxic to the flora and fauna they encountered.[113] In addition, it appeared that some of the tailings ponds were "doing more harm than good."[114] Ponds such as those built on Canyon Creek over old jig tailings rich in unrecovered lead and zinc had raised the water table, causing lead, zinc, and other heavy metals to leach into the groundwater. As the groundwater seeped into the South Fork, that much-burdened stream was once again repolluted with heavy metals.[115]

Settling ponds evidently offered some help with the problem of reducing settleable solids from mining waste water, but offered no real mechanism for reducing heavy metal concentrations. Thus, even after the implementation of modern settling ponds and tailings impoundments, heavy metals continued to contaminate the groundwater and the rivers and streams that flowed from the mining district. Of the approximately five thousand gallons of wastewater that entered Bunker Hill's tailings ponds each minute, two thousand gallons left. Some of the two thousand gallons evaporated but a larger amount found its way into the groundwater system, "carrying with it heavy metals in solution." These heavy metals, once in the groundwater, were difficult to remove.[116]

Another problem Rabe discovered was that the settling ponds were rapidly filling up and mining companies were continually having to buy up additional land for the siting of new ponds. By 1973, for instance, the five ponds dug five years earlier on Canyon Creek by the Star Mines were already nearly full.[117] In the narrow confines of the valley of the South Fork, it was only a matter of time before sites for such ponds ran out. In addition, the filling of the valley floor with a toxic mixture of jig tailings and gravel—a problem that Ellis had noted in 1932—had only worsened over the intervening decades. By the early 1970s this mixture made up the upper eight to ten feet of the valley floor.[118] Every spring during runoff the lower valley was repolluted by the South Fork's fresh reworking of the old tailings.[119] As a solution, Dale Ralston, a hydrologist with the Idaho Bureau of Mines and Geology, proposed channelizing the meandering, heavily braided South Fork through the construction of six-to-seven-foot-high levees. This, Ralston suggested, not only would prevent the river from overflowing its banks every spring and rewashing toxic old tailings downstream but also would eventually succeed in sending tailings from the new riverbed downstream, where they would do less harm.[120]

Although the South Fork saw improvement in the five years following the construction of settling ponds and tailings impoundments, it still remained far from a healthy river. In the summer of 1973 live box tests were performed at eleven sites in the South Fork and one in the relatively pristine North Fork of the Coeur d'Alene. In results remarkably similar to those of the early 1930s, all of the fish placed in boxes in the South Fork below the mining district died within seventy-two hours. The trout placed in boxes adjacent to Smelterville, in the heart of the district, died within five hours. By contrast, the fish put in boxes in the North Fork and in the South Fork above Mullan all did fine. The researchers concluded with understatement that "the South Fork still is a highly unhealthy place for fish."[121]

With regard to the diversity of aquatic insects in the South Fork the story was similar. In 1968, when researchers sampled the insect diversity and quantity in the South Fork at Smelterville, they found absolutely no aquatic insects—a faunal dead zone. Several miles downstream, well below the confluence of the North and South Forks of the Coeur d'Alene, 125 specimens of one individual species were found.[122] By contrast, at a sampling station on the North Fork, 368 specimens of twenty-seven different species were located. In the years immediately following the construction of the settling ponds there was "some detectable improvement in total numbers and species present at a few sampling stations on the South Fork," but researchers found that "the contrast with the North Fork is still striking."[123]

Rabe did not perform an extensive study of the question, but anecdotal evidence indicated that the lower Coeur d'Alene and its lateral lakes also were harmful environments for wildfowl. Idaho Department of Fish and Game wildlife manager Al Bruner reported to Rabe that wildfowl mortality almost always occurred in the spring, with the highest mortality rates generally corresponding with years in which the Post Falls Dam conducted an early fall drawdown of water.[124] Bruner theorized that the early drawdowns left the lakes' mud flats exposed for longer periods of time, permitting higher quantities of their burdens of heavy metals to return to the lakes and their surrounding marshes with the return of water in the spring.[125] Lower Coeur d'Alene River ranch foreman Marion McPeak also reported finding scores of dead and disabled birds on his pastures adjoining the strangely turquoise-colored river.[126] According to McPeak, "In early spring, they fly out of Blue Lake or the river, and land on the lower end of our fields. But they can't fly away again." The ranch foreman reported encouragingly, however, that often if he or his workers could reach the whistling swans and Canada geese in time, pen them up, and feed them uncontaminated grain and water, the birds could be nursed back to health.[127]

The lower Coeur d'Alene's metals-laden waters also took a toll on unwary horses and cattle, especially nursing foals. In the early 1970s, McPeak told a story that encapsulated the danger: "We had twelve mares with nursing foals and put them in this pasture [by the river]. We came down a few days later and found that all the foals had swollen joints. When they moved their knee joints popped like rifle shots. We lost all twelve of them."[128] The ranch foreman also reported having to build new feedlots and pipe in an alternate source of water in order to shield the ranch's cattle from the river's toxic waters.[129]

As it turned out, a type of grass called redtop was a potent factor in the poisoning of horses and cattle along the lower Coeur d'Alene. This grass, common to the area, concentrates heavy metals such as zinc, lead, and copper when they are present in the soil. Redtop was capable of concentrating heavy metals up to four times the quantity they were found in the soil. With soils as contaminated as those found at Cataldo Flats and other areas along the river, redtop's uncanny properties of concentration could prove lethal. Local veterinarian Roy Larson commented, "You cannot pasture a mare with a nursing colt anywhere near the river or the foal will die without drinking the water or eating the [normal] forage." Larson also noted that if a "foal is old enough to be weaned, it 'will die from eating the red top grass.'"[130]

More encouragingly, Rabe found some evidence of the recovery of stocks of aquatic life along the lower Coeur d'Alene since the installment of settling ponds and tailings impoundments in the mining district. By the early 1970s fishermen at Springston, just upstream from the mouth of the Coeur d'Alene River, were catching bullheads, tench, perch, and squawfish. In addition, the stomachs of catfish taken from the lower river in 1973 contained fly and caddis larvae, indicating that, probably for the first time in decades, the river was producing aquatic insects.[131] In another positive sign, Idaho Fish and Game biologists believed that adult cutthroat trout found in Evans Creek, a tributary to a lateral lake, ascended up the Coeur d'Alene River from the lake. They suspected that the fish came up the river during spring, when the high runoff had sufficiently diluted the river water so as to make it habitable. While undoubtedly a step forward from the river's condition in the early 1930s, when Ellis found the lower Coeur d'Alene to be "devoid of life," its status in the early 1970s still represented a huge diminution of the river's productivity since the advent of mining, when fish were observed by the thousands in the big eddy below the Old Mission at Cataldo.[132]

The Health of the Lake

Turning their attention to Lake Coeur d'Alene, Rabe and his students conducted core samples of the lake's bottom sediments to determine the extent of the effects of the mining district on the lake's underwater environment. In results consistent with Ellis's findings, sediments from the underwater delta at Harrison showed zinc, lead, cadmium, and copper to be evenly present to a depth of up to eighteen inches. Although less concentrated than at sampling stations up the river, "sediments of high metal content were found most everywhere at the south end of the lake."[133] Using a small dredge, researchers also sought out aquatic fauna. They found fish-food organisms including fly larvae, aquatic earthworms, and caddis worms, with fly larvae making up the majority of organisms located at the lake bottom. Even the fly larvae, however, became sparse when they came into contact with the heavily polluted delta at Harrison.[134] Using a six-foot core drill sampler to study the lake bottom at Harrison revealed the existence of significant amounts of heavy metals down to a depth of thirty-one and thirty-two inches. Researchers also found that the diversity of tiny algae called diatoms dwindled as the concentrations of heavy metals in the sediments increased and also that diatom shells located in highly metallic sediments often were deformed.[135] The findings, in keeping with those of the early 1930s, indicated that Lake Coeur d'Alene, while more life-sustaining than its tributaries that flowed from the mining district, nonetheless was somewhat ecologically impoverished.

In 1937, in response to the falloff of Lake Coeur d'Alene's cutthroat trout population, the Idaho Department of Fish and Game, devoted to providing a new game fish to Idaho anglers, began stocking kokanee in the lake. Fish and Game biologist Jerry C. Mallet laid the blame for the cutthroat's decline and subsequent failure to thrive in the lake on several main factors. One of these was the long-standing pollution of the Coeur d'Alene River, which denied the species an important spawning habitat.[136] By the early 1970s, one million kokanee were being stocked in Lake Coeur d'Alene each year. Kokanee runs were established on the Saint Joe River in 1973 and Idaho Fish and Game officials wanted to establish runs up the Coeur d'Alene River as part of an effort to replace artificial maintenance of kokanee populations with natural reproduction. The Coeur d'Alene, however, still was not nearly clean enough to support salmon runs.[137]

Enter the Feds

Among the many landmark environmental statutes enacted during the late 1960s and early 1970s were the Clean Air Act of 1970, the amendments to the Federal Water Pollution Control Act of 1972 (the "Clean Water Act"), the National Environmental Policy Act (NEPA) of 1969, and the Endangered Species Act (ESA) of 1973. Of the new federal agencies created to enforce this body of legislation, none would have a greater impact on the environment of the Coeur d'Alenes or on the conduct of business in the mining district than the Environmental Protection Agency (EPA). This agency came into being in December 1970, under the administration of President Richard M. Nixon. The amendments to the Federal Water Pollution Control Act represented an extension of the legislation of 1948 that, importantly, gave enforcement authority to the brand-new EPA. Under Section 402 of the legislation, the EPA had the authority to set and enforce water quality standards for "point source" industrial polluters such as the mining companies of the Coeur d'Alenes. The principal tool granted to the agency was a permitting process under which the EPA could grant industries the right to discharge only certain quantities of pollutants into waterways. Lacking the requisite federal permit, industry would be breaking the law if it released pollutants.

On June 26, 1973, the EPA held a sparsely attended public hearing in the city of Coeur d'Alene. The subject was the issuance of permits to mining companies governing the discharge of wastes into the South Fork of the Coeur d'Alene. At the hearing, EPA officials asserted that restrictions upon the companies' discharges were needed because the dumping of mine wastes and raw sewage into the river had made the South Fork "one of the most seriously polluted rivers in the Northwest." Mining industry representatives in the audience responded that the proposed regulations were "difficult, unreasonable, and in some cases, impossible."[138] Under the new regime the amounts of heavy metals discharged daily into the South Fork were to be cut dramatically: zinc from 9,500 to 1,100 pounds; lead from 540 to 20 pounds; arsenic from 125 to 30 pounds; and suspended solids from 23,000 to 8,000 pounds. In September 1973, waste discharge permits were issued to the Bunker Hill and Sunshine Mining companies.[139] Bunker officials announced that their new wastewater treatment plant would be completed by the end of the year, at a cost of over $1 million and an annual operating budget of $702,000, and that it would cut the release of heavy metals to less than half of the amounts allowed by the EPA permit issued in September 1973.[140] This significant decrease in the release of heavy metal pollutants into the South Fork represented a positive step for one of the most polluted streams in North America, but "Shit Creek," as some locals referred to the river, still was far from healthy.

The early 1980s represented a difficult time for the human populations of the Coeur d'Alenes, as the Bunker Hill Company shuttered its operations, laying off 2,100 workers. Many other mines in the district either curtailed or eliminated production at the time in response to low metal prices and the loss of the nearby smelter. The mining district's downturn, however, meant something different for the waters of the Coeur d'Alene basin and the plants and animals they served. In early 1982, the Bunker Hill smelter ceased spewing its heavy load of lead, zinc, and other toxins into the air for the first time in nearly sixty-five years and the company's zinc plant stopped fouling the South Fork for the first time in over half a century. In 1983, the EPA declared a twenty-one-square-mile area centered on the old smelter complex a federal Superfund site, subject to the new CERCLA legislation (1980) that required corporations, under EPA supervision, to shoulder the burden for cleaning up their environmental messes, inaugurating decades of the agency's large and often controversial role in the affairs of the Coeur d'Alenes.

In the early 1990s, the Idaho Department of Fish and Game received reports of fish in the lower South Fork of the Coeur d'Alene.[141] These were probably the first fish to inhabit the river's waters since the early years of the century. The South Fork remains heavily polluted, however, officially designated as an "impaired river" that requires a remediation plan under Section 303 of the Clean Water Act. Its lower reaches now support reasonably substantial populations of native westslope cutthroat trout, native shorthead sculpin and torrent sculpin, and introduced species of rainbow trout, brook trout, and mountain whitefish, but other sections of the river are far less productive. The middle reaches, particularly the area between Kellogg and Smelterville in the heart of the old smelter district, remain nearly devoid of fish and the benthic macroinvertebrates that local fish rely upon for food. This is likely due to the high levels of zinc that leach into the river from old mine adits and toxic groundwater in the area.[142] Bull trout have not reestablished themselves in the South Fork and westslope cutthroat are rarely found above Smelterville, except in the extreme upper reaches above Mullan. Despite its progress, it appears the South Fork will remain a diminished fishery until its concentrations of heavy metals such as zinc, lead, and cadmium have been further reduced.

The main stem of the Coeur d'Alene and its lateral lakes, in whose lower reaches Rabe witnessed the reintroduction of warm-water fish species in the early 1970s, has continued to make progress as a fishery. Anglers now can hire fishing guides from businesses such as the Castaway Fly Shop in Coeur d'Alene to take them in search of prize catches of westslope cutthroat trout, bull trout, rainbow trout, chinook salmon, bass, northern tench, and tiger muskellunge. All of these species, excepting the westslope cutthroat and the bull trout, have been artificially introduced, and the natives are relatively few and far between.[143] Eating sizable amounts of one's catch, however, is not advisable, as these fish contain relatively high levels of zinc, lead, and cadmium.

The Future of the Lake

While far from certain, the prognosis for Lake Coeur d'Alene may be less favorable than that of its upstream and downstream tributaries. Primarily due to the lake's sheer size, its biological communities never have experienced the level of devastation that its namesake rivers have witnessed, but the lake undoubtedly has served as a sink for the massive heavy metal loads issuing from the upstream mining district. Despite Lake Coeur d'Alene's postcard vistas and sparkling waters that attract thousands of visitors to its shores each year, a monster of sorts lurks beneath its waters. But unlike the legendary monster of Loch Ness, there is no doubt about the existence of the heavy metals that coat the bottom of Lake Coeur d'Alene. Between 350,000 and 440,000 tons of lead coat the lake bottom.[144] Huge amounts of cadmium and zinc also are present in the lake's bottom sediments. The coarser sediments have settled out near the mouth of the Coeur d'Alene River at Harrison, where a twenty-foot-thick delta of metal-enriched sand begins a downward slope of nearly three-fifths of a mile to the lake bottom. A metal-rich sediment layer up to 3.9 feet thick is found near the delta. Finer sediments drift toward the lake's outlet near the city of Coeur d'Alene, where the Spokane River courses toward the city of Spokane, Washington. At the city of Coeur d'Alene the metal-enriched sediments decrease to a thickness of only 4 to 5½ inches.[145] Other than at the delta, the lake's lead concentrations are heaviest at its greater depths, where bottom sediments show a mean concentration of 1,900 milligrams per kilogram (mg/kg) but range up to 7,700 mg/kg.[146]

While for a number of years the prevailing wisdom held that the lead and other heavy metals sitting quietly at Lake Coeur d'Alene's bottom posed no real threat to the ecological health of the lake, that thinking is now being questioned. The lake continues to be polluted with heavy metals from upstream, as is evidenced by the continuing growth of the delta at Harrison. Of the sediments that find their way into the lake, an estimated 80 percent come from the lower Coeur d'Alene basin. The lion's share of these comes from the bed of the Coeur d'Alene River, where many of the old tailings have settled out.[147] While continued pollution by heavy metal sediments is a problem, this does not appear to be the chief threat to the lake's health. After more than a century of intense heavy metal deposition, lead, zinc, and other metals, for better or worse, have become important actors in Lake Coeur d'Alene's complex ecological system.

Short of the removal of the bulk of the metals from the lake's bottom sediments—economically if not technologically unfeasible—it is not clear that the removal of a fraction of the pollution would improve the ecological condition of the lake. One of the chief hazards to the lake, in fact, may derive from lowered deposits of zinc due to the cleanup of the lateral lakes area of the lower Coeur d'Alene. Zinc, it happens, plays an important role in shaping the lake's chemistry. Binding with plant matter, and, through its toxic properties, keeping aquatic plant growth in check, the heavy metal plays an important role in holding down the lake's plant populations.[148] But there are indications that zinc's heroic efforts are not keeping pace. A US Geological Survey (USGS) limnological study of the lake in 2004–2006 found increased concentrations of nutrients such as phosphorus and nitrogen and thus increased lake productivity of plants.[149] Aquatic plants and algae grow better in such nutrient-rich environments. When these organisms die and decompose, oxygen dissolved in the water is consumed. A danger exists that with a decrease in the quantity of dissolved oxygen in the water, chemical processes will be set in motion that lead to a heightened rate of release of the heavy metals that are present in lake-bottom sediments. Such a process not only would produce a more toxic lake for fish and humans, it also would increase the lake's rate of eutrophication, as additional algae growth would be encouraged.[150]

While zinc currently is keeping the lake's main algae community, diatoms, in check, it is feared that with the increased nutrient loading of the lake communities of zinc-resistant green algae and bloom-forming species of blue-greens will take root. This self-perpetuating process has the potential to, in the tempered words of the USGS, "increase risks to human health and adversely affect the environment and the economic and social welfare of the community."[151] Simply put, the combination of eutrophication and the release of heavy metals from bottom sediments into the water stream has the potential to result in a severely degraded lake. The USGS is not the only scientific voice to express deep concern about the future of the lake. In its 2005 report the National Research Council concluded:

> It is also unclear what will happen in Lake Coeur d'Alene. . . . The major question is whether the lake will become more eutrophic, and, if so, what effects this will have on the lake's chemistry and biota. There is substantial concern that changes in the lake's chemistry could result . . . in the release of contaminants currently bound in the sediments coating the lake bottom. These released contaminants could be toxic to fish and other aquatic biota and, therefore, in conjunction with the other effects of eutrophication, could cause significant changes in the lake's biological systems.[152]

It turns out that the hundreds of thousands of tons of heavy metals once thought to be safely encased in Lake Coeur d'Alene's bottom sediments are significantly more volatile and less predictable than was once believed. Instead of being locked up tight, they evidently are chemically available under certain circumstances. Ecological ironies abound. Pollution in the form of heavy metals such as zinc that once was able to protect the lake's health may become unable to do so any longer. Efforts to clean up heavy-metal pollution in the lower Coeur d'Alene River, by decreasing zinc releases, may prove profoundly harmful to Lake Coeur d'Alene. One is left to wonder about the prospect of having to add zinc to the lake in order to "remedy the ecological problems" of an environment already suffering from contamination by heavy metals.[153] One thing seems sure; there is no returning to a "pristine environment" in the Coeur d'Alenes.

In the spring of years of high runoff, it is not uncommon to see a discolored plume of water riding atop Lake Coeur d'Alene as it snakes its way from the mouth of the Coeur d'Alene River toward the lake's outlet at the city of Coeur d'Alene. Spring runoffs constitute a principal means by which the heavy-metal-laden sediments of the Coeur d'Alene River system are transported into the Spokane River, which flows to the city of Spokane.[154] Although concentrations of the usual congeries of the Coeur d'Alenes' metals (e.g., zinc, lead, arsenic) are somewhat dissipated by the time they reach the waters, riverbanks, and beaches of the Spokane River, they still are often found at levels considered unsafe for humans and animals.[155]

It is hard to believe that the early mining chieftains of the Coeur d'Alenes—men such as Bunker Hill's F. W. Bradley and Stanly Easton and Hecla's Harry Day—ever imagined that one day in the early twenty-first century a federal pollution-control agency would be hauling giant shovel-loads of heavy-metals-laden dirt and sand away from popular Spokane-area beaches. Although men such as Bradley and Day undoubtedly understood that tailings from their operations invariably made their way down the South Fork to the Coeur d'Alene River, the notion that their metallic deposits, found in the sands of the Spokane beaches, would constitute such a threat to human and animal health as to warrant the action of a federal agency to remove them, likely would have been inconceivable to them. And yet, just this scenario has been played out repeatedly over recent years, as popular beaches along the Spokane River have been "remediated" by the EPA and state agencies.

Late nineteenth- and early twentieth-century mining industry chieftains like F. W. Bradley did not, as a rule, put much thought or energy into questions involving the ecological impact of their operations. Busy applying the principles of Manifest Destiny to the industrial conquest of Nature for the rewards offered by capitalism, and proud of their role as pioneering agents of progress, as a class they appear to have fretted little about the social costs—whether in workers' lives or lifeless rivers—of their endeavors. Like other products of the "heroic age of mining engineering" such as John Hays Hammond and Herbert Hoover, Bradley and Easton almost certainly viewed themselves as heroic figures: engineers and businessmen who, having carved an industrial empire out of the Idaho wilderness, provided vast quantities of relatively inexpensive lead, silver, and zinc to a rapidly expanding nation of industrial producers and consumers.[156] As turn-of-the-twentieth-century westerners, they likely viewed the region as vast and sublime—with rivers, lakes, and airstreams acting as inexhaustible sinks capable of handling nearly any amount of waste by-product of industrial processes—and its resources as virtually inexhaustible. Easton was an avid hunter and fisherman who, with no apparent sense of irony, served a term as a member of the Idaho Board of Fish and Game. By the year 2001 their mining district's tailings, if set inside a football field, would have reached four miles into the air, but this fact probably would have filled Bradley and Easton more with a sense of pride than with ecological concern.[157]

The generation of empire builders who directed the creation of a spectacularly productive lead, silver, and zinc mining and processing complex in the far reaches of northern Idaho is long gone. Much honored in their time, names such as Bradley, Easton, and Day now are largely forgotten, even in the Coeur d'Alenes. Their legacy is very much alive, however, in the communities, the wealth, and the industrial products that their efforts helped to spawn. That legacy too lies embedded in the millions of tons of heavy-metal sediments that continue to confront the Coeur d'Alene watershed, its plants and animals, and its people with significant health and environmental costs and problems, and likely will do so for many decades to come.

CHAPTER 4

FOUL HUMOURS

Just in time for the 1917 Independence Day celebration, a major lead smelter was born west of Kellogg. Eleven years later, a younger sibling, a zinc refinery, joined the metal-processing complex. At the time of their completion and for some years to come, these custom facilities were praised as technological marvels that contributed to the economic development and general progress of northern Idaho and the Inland Empire. The facilities were bathed in the light of what philosopher Marshall Berman has called "the modern romance of construction."[1] Nearby residents, however, soon complained of the acrid, vegetation-killing sulfur dioxide (SO_2) smoke the Bunker Hill plants produced. By the 1950s and 1960s these complaints had become louder and more insistent, particularly from federal agencies such as the US Forest Service. In response, the company began to implement technological changes, such as the construction of a sulfuric acid plant, designed to bring its sulfur dioxide problem under control. With the passage of the Clean Air Act in 1970, more stringent controls were placed upon the company's sulfur dioxide emissions, standards Bunker Hill struggled to meet until its construction of the "tall stacks" in 1977. For the company, the 1970s bore out the truth of the observation that "socially recognized risks"—in the case of the Coeur d'Alenes, the risk to the health of humans and vegetation from sulfur dioxide—"contain a peculiar political explosive; what *was* until now *considered unpolitical becomes political—the elimination of the causes in the industrialization process itself.* Suddenly the public and politics extend their rule into the private sphere of plant management—into product planning and technical equipment."[2] Also in the 1970s, Bunker Hill, from motives that mixed genuine ecological concern with a desire to curry favorable public opinion following the self-inflicted publicity damage of the 1973–1974 lead health epidemic, embarked upon a major revegetation effort in the Silver Valley. This effort offers a prelude to more successful EPA efforts in the late 1990s. This chapter follows several decades of concern and wrangling over the Kellogg area's air quality and examines the costs of an unfettered economic activity to local environments and landscapes.[3]

Genesis: A Smelter Is Born

The health and environmental concerns that later would surround Bunker Hill's smelter were scarcely in evidence at the facility's unveiling. Ground had been broken for the company's lead and silver smelting and refining works in June 1916. On July 5, 1917, furnace number one was blown in and the smelter opened to considerable local and some national fanfare. Writing in the lead mining trade publication, the *Engineering and Mining Journal,* on the eve of the smelter's inauguration, Idaho's state inspector of mines Robert N. Bell referred to the event as "an epoch-making occurrence in the history of this remarkable smelting-ore district."[4] Bell praised the company's siting and the efficiency of the plant, calling it "elaborately planned for transportation and convenience, gravity handling, and supplied with every modern mechanical device obtainable."[5]

The million-dollar, state-of-the-art facility, one mile to the west of the tunnel into the Bunker Hill Mine, boasted four Dwight & Lloyd roasters, a twenty-six-foot wedge furnace, and three steel-jacketed blast furnaces.[6] Its superstructure consisted of structural steel, cast iron, and red brick. Liquid lead bullion was to be transferred first to a drossing plant and then to the new refinery, which contained two softening furnaces, four desilverizing kettles, two refining furnaces, and four merchant kettles. The refinery's silver division contained six retorts, two cuppelling furnaces, one fine-silver furnace, and a crucible furnace for good bullion.[7] Waste gases were to be dealt with via a baghouse and a concrete stack measuring two hundred feet in height by twelve feet in diameter.[8]

The construction of its own smelting complex accomplished at least two major goals for Bunker Hill. It enabled the company to free itself from dependence upon the Guggenheims' virtual stranglehold on the lead smelting industry while also making Bunker one of the few lead-silver mining firms in the country capable of smelting its own product along with the ores of others.[9] For years Bunker had shipped its ore to the Guggenheims' AS&R smelter at Tacoma, Washington. Company managers F. W. Bradley and Stanly Easton had long nursed resentments and fears over what they viewed as ASARCO's capacity to capriciously raise smelting rates. With the construction of their own works, the company could leave those fears behind.

After considering several sites in the Northwest, including ones in Tacoma and Portland, Bunker decided to build its new plant right in its own backyard. Principal factors behind the siting decision were: the belief that the relatively sparse, economically dependent population of the Kellogg area was less likely to oppose a lead smelter's noxious effluent than were such populous locales as Tacoma and

FIGURE 4.1. Bunker Hill and Sullivan Mining and Concentrating Company lead smelter, October 1917.

Portland; the easy, efficient access to its own and other district ores; the fact that Bunker Hill already owned the land upon which it wanted to develop the smelter; and the relatively friendly treatment that the company—locally referred to as "Uncle Bunker"—could expect to receive from the area's populace generally and local government officials more specifically. Not surprisingly, the company did not stress pollution issues nor the relatively benign nature of the taxation and regulatory approach of local government officials when explaining its decision to locate the smelter near Kellogg, instead highlighting factors of economy in construction and operation.[10]

But smoke pollution concerns certainly were on the minds of Bunker Hill executives when they made the decision to construct the smelter one mile west of Kellogg, where the town of Smelterville soon took shape. Bunker Hill executives were well aware of the various "smoke lawsuits"—on behalf of both private citizens and the US Forest Service—that had beset the Anaconda Company and other members of the US smelting industry since the first years of the new century. In order to buffer the company against this threat, Easton began purchasing "smoke damage" easements from local landowners before the decision to site the smelter was announced.[11] These easements, like the "water damage" easements bought from farmers bordering the Coeur d'Alene River, essentially purchased a right to engage in a certain form of pollution on a landholder's property for an indefinite period of time. The company also spent considerable sums on pollution-control equipment; namely, the baghouse and a new, relatively revolutionary piece of technology, the

Cottrell electrostatic precipitator.[12] Developed by University of California, Berkeley, chemist Frederick Cottrell, the precipitator employed a powerful electrostatic charge to separate out toxins such as lead and sulfur oxide from the waste fumes of a smelter. If installed and used correctly, Cottrell's invention represented a significant advance in smelter pollution control, but it was far from a panacea. In an effort to forestall any potential complaints from its large land-managing neighbor, the US Forest Service, Bunker sought and received assurances from the federal government that it would not be sued for damages to nearby forest land in return for a pledge to pay for any damages it might inflict.[13]

In its earlier efforts at purchasing easements from disgruntled downstream landowners, Bunker Hill had enjoyed the support of its fellow Coeur d'Alenes mine owners, in the form of the MOA. In this case, however, as the sole smelter owner in the district, Bunker would absorb all the cost and liability. Adopting a two-pronged strategy similar to the one it had employed to deal with its water pollution problems, Bunker on the one hand resolutely fought any and all smoke damage lawsuits, determined not to give ground on the economically vital terrain of legal liability. On the other hand, however, the company was willing to purchase pollution easements or buy land outright from those it deemed to be in the pathway of its smelter smoke.[14] Bunker was even willing to pay out what it considered to be reasonable sums in cases where it believed there to have been "smoke damage" to someone's land. The company additionally was prepared to spend considerable sums on pollution-abatement devices—examples include the Cottrell precipitator, an expansion of the baghouse in 1923, and the suction dredge at Cataldo—but there were limits, as was shown with its refusal to slurry its tailings to an impoundment pond. Bunker Hill proved willing to swallow the relatively predictable, if not always small, costs of keeping up with industry standards for pollution-control technology and by "paying to pollute," in the form of buying land and pollution easements from neighboring property owners. Viewing the possibility of legal precedents holding it liable for damages for harm done by the release of its pollutants as anathema, however, the company pursued a policy of aggressively defending itself in court and never accepting legal liability for harm that might have resulted from the effluents of its works.

As befit a large, professionally run, twentieth-century corporation, Bunker Hill developed regular procedures for handling what the company referred to as its "smoke damage" program. This program included the purchase of smoke easements, direct purchases of land, and the payment of damage awards. The company developed legal forms to secure certain, specified rights to engage in the air pollution of others' lands—for instance, with smoke from its smelter, on identified tracts of land—while also specifying that landowners would forgo their right to sue the company for pollution-related damages. Professionals were put on retainer to investigate claims and otherwise administer the program—a veterinarian, a pathologist, a mechanic, attorneys, and someone to "cruise timber."[15] The sums expended on the program suggest that keeping air pollution lawsuits to a minimum and maintaining good community relations were priorities to the company. In the eighteen years spanning December 1, 1936, through November 30, 1954, the company spent close to $394,000 on its smoke damage program, or an average of $21,888 per year. The lowest expenditure, $10,165.35, came in 1940, whereas the highest payouts occurred in 1950, when $36,390.61 was spent to assuage the damage. A portion of this money went to administering the program, of course, but significant sums went directly to landowners in the form of pollution easements, ranging from a low of about 25 percent to a high of roughly half the total expenditure on the program for the period examined here. Payments for "crop damage" or "crop and livestock damage," on the other hand, constituted only minute portions of total expenditures—from nothing in 1942 and 1943 to a high of $365 in 1937. The company presumably preferred the relatively predictable and tidy nature of easements and land purchases over the more problematic business of making post facto payments for damage claims.

Additionally, during the dozen years running from 1928 to 1939, Bunker appears to have spent $70,501.28 on its smoke damage program. In an internal company memo, a Bunker official who went by the initials "C.P." made little pretense about the fundamentally defensive nature of these expenditures or about the condition of the lands purchased, writing that "the Smelter has also spent, up to and including the year 1939, $25,822.07 for the purchase of smoke damaged land. . . . This land is farm land and timber land that has been rendered valueless because of smoke damage and is of no use to the Company except as protection against smoke damage claims."[16] Following over twenty years of lead smelter and a dozen years of zinc plant operations, C.P. noted, "There is some timber on the land purchased, most of which is second [growth] and a large part of which has been smoke killed." By 1940, the company evidently owned thirty-nine "smoke-damaged" properties, totaling 7,831.98 acres, on which it paid local property taxes.[17]

The pathology contract with Bunker Hill was relatively lucrative, as was to a lesser extent the veterinary contract. From 1937 to 1943, Bunker paid the pathologist between $3,828.31 and $4,818.10 annually for his services, not including the company's annual payment for his supplies.[18] The veterinarian's annual payment ranged from $1,183.70 to $1,994.29. In 1949, a new itemized expenditure line entitled "soil erosion" appeared for the first time in company records. A $1,372.26 payment was made. Bunker Hill's electrolytic zinc smelter, which began production in 1928, put out lots of sulfur dioxide. This sulfur dioxide caused severe damage and often death to area pine trees. Damaged and dead trees could not hold soil in place as they once had, especially on the steep-sloped mountainsides that surround the Kellogg area. It appears that, by the late 1940s, in addition to its other "smoke damage" payments, the company was compensating local property owners for damage to hillside soils.

A Zinc Plant for Kellogg

Once viewed as nearly worthless, by the 1920s considerable quantities of Coeur d'Alenes zinc was being sent outside the district for processing. Enjoying success in their newfound role as a custom smelter of Coeur d'Alenes lead and silver ores, Bradley and Easton decided to build a plant to process the area's valuable zinc resources as well. This decision likely was spurred by the company's 1922 purchase, along with its partner the Hecla Mining Company, of the valuable Star Mine in Burke. Star Mine ores, like others in the Pine Creek drainage, contained large amounts of zinc along with lead and silver.[19] The Bunker Hill smelter was unequipped to treat the zinc portion of these ores. While encountering difficulties in experimental efforts to smelt the Star Mine zinc ores, Bradley heard about a method that had enjoyed some success in the Anaconda and Trail, British Columbia, smelters. The Tainton process, developed by U. C. Tainton, employed sulfuric acid to leach zinc ore and then an electrolytic process to extract zinc from the zinc sulfate solution.[20] After experiencing success with the Tainton process, Bradley became determined to secure exclusive rights to use it for his company and thus garner Bunker Hill an important competitive advantage. By agreeing to pay Tainton a 10 percent royalty, in May 1922 he did exactly that.[21] When it went online in October 1928 near Government Gulch, four miles west of Kellogg, Bunker Hill's electrolytic zinc smelter set a new industry standard by producing at a level of purity theretofore unseen: 99.99 percent pure zinc.[22] The smelter represented a $2 million investment on the part of Bunker Hill and Hecla. Once up to full production speed, the new plant was expected to produce sixty tons of zinc per day.[23]

While its zinc might have been pure, the plant could hardly claim the same for its waste gases. Whereas the principal environmental burdens produced by Bunker's lead-silver smelter were heavy-metal oxides such as lead, cadmium, and arsenic, the zinc plant's chief offender was sulfur dioxide (SO_2).[24] As scientific studies of pine and fir trees, and other vegetation near lead, zinc, and copper smelters had shown, airborne sulfur dioxide in sufficient concentrations could have devastating effects. In a study in 1903–1904, J. K. Haywood, chief chemist of the Miscellaneous Division of the US Bureau of Chemistry, determined that sulfur dioxide and trioxide from the Keswick smelter near Redding, California, were damaging and destroying vegetation. Haywood's report, published in 1905, concluded that sulfur dioxide "when present in very minute amount kills vegetation," and that vegetation as distant as nine miles from the smelter had been "greatly injured" by sulfurous effluent.[25] Haywood's subsequent investigations of the conditions surrounding smelters in Ducktown, Tennessee, and Anaconda, Montana, only confirmed his earlier conclusions. Examining the effects of the arsenic- and sulfur-rich effluent of the Anaconda Company's giant copper smelter in 1906–1907, Haywood determined that tree regeneration had been impaired and vegetation injured in a large radius extending from the works—for at least ten miles to the north, six miles to the south, thirteen miles to the west, and fourteen miles to the east. Haywood found differing resistances to the smoke among different species of trees. While junipers proved almost impervious, red firs were highly susceptible. Lodgepole pines were found to be less susceptible than red firs, but still sustained injury at distances as great as ten miles from the smelter.[26] While large quantities of arsenic were present in the copper smelter's smoke, Haywood left no doubt that sulfur dioxide constituted the principal cause of harm to area vegetation.[27]

Later scientific research built upon and reinforced the findings of Haywood and others. In a major international dispute that lasted from the late 1920s through the early 1940s, a group of northeastern Washington farmers alleged that fumes from the Consolidated Mining and Smelting Company of Canada's lead-zinc smelter, located along the upper Columbia River at Trail, British Columbia, had damaged their crops, trees, and livestock. In work that was readily accepted by the scientific community, George G. Hedgcock, the senior US Forest Service pathologist, after first distinguishing smoke damage from drought and winter damage to Douglas fir trees, ponderosa pine, and forest shrubs, proceeded to map the geographical distribution of sulfur dioxide injury in three distinct zones that were based upon the percentage of smoke-damaged trees that the areas contained. Hedgcock's work showed that while the worst and most widespread damage generally occurred closest to the smelter, tree damage extended as far as forty-nine miles down the Columbia River valley.[28]

It is likely no coincidence that Bunker Hill began spending more heavily on its "smoke damage program" in 1928, the year its zinc plant began operating. The company was not content, however, merely to purchase quiescence from its private-sector neighbors. Following the precedent set by the Anaconda Company, which successfully swapped thousands of acres of forest lands it owned for Deer Lodge National Forest lands that either already had been, or likely would be, damaged by its smelter smoke, Bunker began swapping some of its large forest holdings for Coeur d'Alene National Forest lands that lay in the pathway of smoke from its zinc and lead refineries.[29] By obtaining ownership of national forest lands in the pathway of their smelter smoke, Bunker Hill seemingly ensured the continued right to pollute these lands without fear of legal recrimination from the federal Department of Justice.[30]

Despite Bunker's previous agreement with the forest service and its continued efforts to purchase peaceable relations with the local population, in the immediate postwar era (1946–1950) heightened complaints about the zinc and lead plants' effects on area vegetation pushed the company to take pollution-control measures. In 1946, Bunker Hill and the Coeur d'Alenes' other major mining companies approached the US Forest Service about the prospect of securing an annual sustained-yield cut of timber from the neighboring Coeur d'Alene National Forest.[31] The companies annually consumed tremendous quantities of timber, principally Douglas fir and western larch—twenty-two million board feet in normal years of good production—primarily to replace old mine timbers that supported thousands of miles of underground tunnels.[32] Facing a dramatic shortage of available timber, the mining companies turned to local forest service officials for help. The companies wanted the forest service to set aside sections of the neighboring Coeur d'Alene National Forest for their exclusive use and guarantee approximately twenty million of the forest's annual thirty-three million board feet cut for their use.[33] Perhaps sensing a rare opportunity, forest service officials raised the subject of the local lead and zinc refineries' sulfur dioxide output and its effect on area tree growth and forest health. Regional Forester P. D. Hanson, noting his department's efforts to improve its management of the forest resource so that sustained timber yields of the type the mining companies were requesting could be maintained, suggested that the mining companies begin shouldering their portion of the burden for local forest health.

Internal Region One documents from 1946 reveal that, in fact, by this date forest service officials harbored significant concerns about the effect of Bunker Hill's smelting and refining operations on both area forests and the local ecology.[34] Discussing how best to frame the upcoming discussion over guaranteed timber harvests, Coeur d'Alene National Forest supervisor K. A. Klehm suggested to his superior, Hanson, that they should base their presentation upon "a factual account of the merchantable timber on national forest land within the mining zone," and "show the necessity for increasing the productivity of other forest lands in the area and should suggest correction of the smelter fume damage."[35] In a rough draft of a letter to Bunker Hill assistant secretary Ira A. Robson, Regional Forester Hanson expressed three main points: alarm over the increasing levels of sulfur dioxide damage on forest lands affected by refinery smoke, concern over heightened soil erosion and fire danger on impaired forest lands, and the irony of the fact that the very mining companies asking the forest service to guarantee them substantial annual timber harvests from the Coeur d'Alene National Forest were also the parties most responsible for the destruction of timber resources in that forest.[36] In the draft, Hanson mentioned that on a recent tour of the forest's "smoke damaged areas," he beheld "some rather startling erosion problems" and added, "We recognize that erosion is becoming an ever important and increasing problem in the fume-damaged area."[37] Trees killed by Bunker Hill's sulfur-dioxide could of course no longer serve to prevent the erosion of soil on the steep hillsides of the mountains that framed the narrow valley.

Turning to another hazard posed by the area's many killed (but still standing) trees and bushes, Hanson wrote, "we also realize that the killed material is also greatly increasing the fire hazard in the area."[38] No one needed to remind residents of the Coeur d'Alenes of the acute dangers posed by forest fire—the narrow, steep-walled valleys of the area were surrounded by forested mountains and the massive complex of wildfires known as "the Big Burn" had galloped through Wallace in 1910, destroying almost half the town. In 1945 the Coeur d'Alenes again were beset by a major wildfire. Turning to an irony of the mining companies' request, Hanson wrote, "This damage to timber [from refinery fumes] is a little difficult to reconcile with the urgent demand on the part of the mining industry for additional forest products."[39] Another irony lay in the fact that had it not destroyed the timber on much of its own forest lands, along with that lying on neighboring forest service territory, Bunker Hill would have been in a far better position to meet its current demand for wood. Thus Hanson wrote, "The area tributary to Kellogg, if in production, would no doubt in itself contribute materially to supplying the wood products that are needed to operate your mine."[40] Mildly chiding Bunker, Hanson mentioned that since its fume damage was "expanding rather rapidly," the company might want to take "some control measures, in addition to those now exercised," in order to stop destroying so much of the timber on its own lands.[41]

In 1950, responding to demand generated by the expanding postwar economy, Bunker's lead smelter produced at record levels and the zinc plant closely followed suit. Increased smelting activity coupled with unchanged levels of pollution control, however, invariably resulted in the release of greater quantities of toxic waste gases such as sulfur dioxide into the airstream. In response, in 1950 local forest service officials increased their complaints to the company over the loss of trees on national forest lands adjacent to the smelter and the zinc plant.[42] At the behest of the USFS, the company began construction of a sulfuric acid plant as an addition to the zinc smelter.[43]

Completed in April 1952, the plant turned a harmful waste product, sulfur, into a commercial product, sulfuric acid, used in making fertilizer. The company soon was producing one hundred tons of acid a day—acid that no longer rained down on the trees and people of the Silver Valley.[44] The company initially struggled to find markets for all of its sulfuric acid, however. In the view of Stanly Easton, this difficulty arose "because of the remoteness of the plants," but undoubtedly the situation also owed to the fact that at the time a number of other US smelters were adopting the same approach to their sulfur dioxide pollution problems.[45] The predictable result was a glut on the sulfuric acid market.[46] Although the sale of its sulfuric acid would always be problematic for Bunker, the challenge was lessened in late 1956 when the J. R. Simplot Corporation, an Idaho fertilizer manufacturer, agreed to purchase 80 percent of the capacity of Bunker's acid plant for the next eleven years.[47]

Domestic consumption of lead rebounded from Great Depression–era lows during World War II and the immediate postwar years. In 1932, during the depths of the Depression, domestic lead consumption dropped to 396,000 metric tons (t) but by 1947 that figure had nearly tripled to 1,090,000 (t) and by 1953 consumption had ballooned to 1,460,000 (t).[48] On the heels of a record production year at its lead smelter, Bunker Hill decided to make a major, $3 million expansion of the lead smelter that would permit the facility to smelt up to one hundred thousand tons of ore per year,[49] a figure representing just over 23 percent of primary US lead production for 1951.[50] The expanded and modernized facility went fully online by the end of 1953.[51] The heightened smelter capacity forced Bunker for the first time in its history to look outside the Coeur d'Alene mining district for the procurement of significant quantities of ore to smelt.[52] With hindsight, however, it appears the smelter expansion may have been ill advised. As early as the late 1950s the company struggled to secure enough concentrates to run the smelter at capacity. In the late 1960s Bunker needed to procure about 60 percent of its concentrates from outside its own mines; in the mid-1970s that figure was 50 percent, and by 1981 it had reached 75 percent.[53] The company received much of this ore from other companies' mines within the district, but it also was forced to procure concentrates from as far away as Peru.[54]

During the second half of the 1960s the Bunker Hill Company became noticeably more concerned about dealing with its sulfur dioxide problems. Desiring increased revenues from sulfuric acid sales, faced with a local population that had become less tolerant of air tainted by the acrid, foul-tasting, and foul-smelling substance, and challenged by new federal and state standards for air pollution, the company focused significant resources on sulfur-control technology.[55] In September 1965 Bunker announced its plan to spend $8 million over the next two years to expand its zinc plant and double the capacity of its sulfuric acid facilities.[56] In 1970 the company spent $6.5 million to purchase and install a German-made sintering machine and a new baghouse at its lead smelter.[57] The German technology, a Lurgi Updraft Sinter Machine, was designed to remove sulfur from gas before it was sent to the baghouse for further cleaning.

Covering the story was a daily newspaper from the Inland Empire's hub, Spokane, Washington, a city of 170,000 that had prospered greatly through its ties to the mining and smelting, logging, and farming activities of a wide hinterland.[58] In its article on the new sintering machine, the *Spokesman-Review* noted environmental tradeoffs that were fundamental to the eighty-five years of industrial modernity in the Coeur d'Alenes—while mining and smelting had netted $2 billion in profits, "the costs in terms of a deteriorating environment have been high—the Coeur d'Alene River polluted and the hillsides browned as acid-laden smog killed trees and shrubbery."[59] Bunker Hill officials struck a different tone, however, promising visions of a green future for the mining district. Perhaps overcome with visions of the technological prowess of its new pollution-control systems, the company forecasted that the valley would be green with new plant life within five years. More realistic were the predictions of Robert Montgomery, chairman of the Idaho State Air Pollution Commission. While acknowledging that the installation of the Lurgi had enabled Bunker to meet current state air standards, Montgomery projected that the state likely would toughen its standards in a few years and the company once again would find itself out of compliance.[60]

Despite the outward confidence displayed at the unveiling of the Lurgi machine, at least some in the company were aware of the looming challenges posed by what promised to be an increasingly strict arena of environmental regulations. Gene Baker, Bunker's chief of environmental affairs, framed the situation in a May 1970 in-house talk in the following manner, "As you all know, there has been a lot of emphasis lately on environmental control. A lot of pressure at various points—mostly against industry. This pressure continually changes but out of it are evolving some firm courses which affect us and although we may not agree—they're there and with force."[61] Baker noted the company's role in harming local vegetation and possibly human health with its sulfur dioxide emissions, stating, "Our emissions have destroyed most of the natural vegetation in the valley and although there is no proof that our SO_2 is damaging to health, there is more and more evidence to support the belief that it is."[62] Evidently more concerned with protecting his company than with the preservation of local flora and fauna, Baker advised his colleagues on how best to approach the stormy sea of looming environmental regulation. Noting that while the new sintering machine would "give us the capability" to meet Idaho's sulfur dioxide requirements established in 1968, recent federal recommendations were far more stringent than the state standards.[63] Predicting that Idaho soon would revise its rules to bring them more in line with federal guidelines, Baker suggested that the company make efforts to keep the Idaho rules relatively weak. Observing that Bunker's "success in obtaining the most lenient regulations" was dependent upon "having satisfied local citizens" and also on "having an air pollution commission which is sympathetic to industry," Baker targeted upcoming local public hearings on air quality in 1971 as the critical time and place for the company to make its case for relatively lenient state sulfur dioxide standards.[64]

Observing that "the local citizens are becoming less and less tolerant" of sulfur dioxide levels, Baker advised that the company use its new sintering machine and sulfuric acid plant to full capacity in the coming year and find markets for all of the resultant sulfuric acid. Only by getting maximum production from its new sulfur-removing technologies could Bunker reduce "SO_2 emissions to currently acceptable levels which will restore our credibility with valley people."[65] But there was the rub. At that point Bunker Hill had a market for only 55 percent of its sulfuric acid. Unwilling to take a loss on excess sulfuric acid, when faced with this situation previously the company had left portions of its acid plant idle. As Baker well understood, the company remained highly reluctant to take a loss on acid it could not sell. Thus he and Bunker Hill president Frank Woodruff both stressed the need to "somehow market all of the acid."[66] Unlike its attitude toward purchasing smoke and water easements, Bunker evidently was unwilling to view the production of unsalable sulfuric acid as an acceptable price to pay for the maintenance of good local public relations, let alone a cleaner and more healthful Silver Valley environment. Baker nonetheless recited to his colleagues the potentially dire consequences of failing to fully utilize the new sulfur collection systems, noting that, "If we fail to meet this commitment, we will be condemned by the public and will be in violation of State standards. In this eventuality, we would face possible legal action, and might be forced to curtail operations."[67]

Citing air pollution regulations that had recently been adopted by the State of Washington mandating that its Air Pollution Control Board "obtain and maintain the cleanest air possible consistent with the highest and best practicable control technology," Baker cautioned that this type of approach was likely to be adopted by all the states.[68] The environmental officer also warned that to keep up with likely mandates to employ "best practicable control technology" to address its issues with sulfur dioxide and aggregate particulate matter, the company would have to spend considerably on technology and upgrades. Baker's preliminary five-year plan predicted the need for $1,482,000 in environmental expenditures. Of this, roughly $1 million was to be directed toward air pollution control efforts, with projected expenditures of $690,000 at the lead smelter and $272,000 at the zinc plant, or roughly a 40 to 50 percent increase in company spending on pollution controls over the previous five-year period.[69]

On March 3, 1971, the Idaho Air Pollution Commission held a public hearing in Kellogg. The subject was Gulf-Bunker's request for a three-year delay in the implementation of the state's 1968 sulfur dioxide standards. The company, having already been granted extensions on compliance to the present date, now explained that a further extension was required due to fluctuations in the sulfuric acid market that prevented it from selling its excess acid. A number of local residents took the opportunity to testify before the state commission. Opinions varied strongly on an issue that pitted the economy, local jobs, and company loyalty against personal health and environmental quality. An undercurrent of anti-Gulf sentiment at the hearing suggested that the new company owners had yet to win the local loyalty that had long been Bunker Hill's to command. N. C. Perkins, manager of a local bank and past Chamber of Commerce president, expressing the view of those who feared that denying the extension would mean the slowdown or outright closure of the company's local operations, stated:

> I think the seriousness of this is quite evident to each and every one of us that is here and that should the variance not be granted, the economic disaster that could happen . . . I don't believe can be over emphasized. . . . We all depend, whether we know it or not . . . upon this industry for our livelihood. If there is no way of living here, if we are not going to be able to work here, the coming generation is not even going to be able to stay here.[70]

Some of those testifying in favor of the company's request for an extension referred to what they viewed as Gulf-Bunker's recent good faith effort to comply with air quality regulations—the expenditure of $6 million on pollution-control technology including the recent purchase and installation of the Lurgi machine—while entreating the commission to grant the company needed time to figure out how to market its excess acid. Demonstrating its continuing support for the company, the Kellogg City Council, represented at the hearing by Mayor Roger Fulton, went on record as supporting Gulf-Bunker's extension request. A principal factor in the council's unanimous resolution was "the economic impact that would be imposed on the City of Kellogg in revenue loss if the Bunker Hill Company operations were curtailed or shut down."[71]

Opposition to the extension request came from, among others, Kellogg doctors, sufferers of respiratory disease, and the student body of Kellogg High School. A local resident and practicing physician, Dr. Keith Dahlberg, cast doubt on the integrity of the company's claims that, given more time, it would handle its sulfur dioxide situation. Dahlberg noted that the company previously had made repeated promises that it would take care of its sulfur dioxide problem, first claiming that it would do so by late 1969 and later pushing the date to January 1971. Turning to the health aspects of the situation, Dahlberg testified that while sulfur dioxide itself is not harmful to humans in concentrations of less than three to five parts per million, in the Silver Valley people were breathing in not pure sulfur dioxide but "conversion products of combined SO_2, fog, metallic oxides, and other particulate matter that we find in our local smog." Noting that little was known about such "conversion products," the doctor suggested that they might be more harmful to human health than pure sulfur dioxide. Speaking directly from experience as a local caregiver, Dahlberg testified with even more certainty, stating, "But [for] those who have some respiratory diseases, there is no question that the disease is made worse on days in which the smelter smoke is strong. . . . Certainly something is affecting our asthma and emphysema patients, which appears simultaneously with SO_2 smog, and which does not appear to affect them in its absence." The doctor cited the case of an asthma patient whom, the previous March, he had felt compelled to discharge from West Shoshone General Hospital before she had received care, "in order to get her out of Kellogg's air," since "she simply could not breathe it."[72]

Dahlberg, like others opposed to granting the extension, also charged that Bunker executives, along with many other people, chose not to live in Kellogg due to its bad air. Noting that Kellogg's population had declined by two thousand over the past decade, the doctor stated that "local business suffers as more and more people move elsewhere to live, because of our air." Turning to the local scenery, Dahlberg indicated that Kellogg was not yet the verdant landscape envisioned by Bunker executives the previous fall, saying, "The effect of sulfur dioxide on the local vegetation is obvious to anyone who has seen our yellow lawns and hillsides bare of evergreens." Dahlberg also questioned Bunker's strategy of harnessing its air quality program to what he described as the dubious economic plan of marketing all of its potential sulfuric acid on the oft-glutted sulfuric acid market. Instead of "putting all their eggs in one basket," the doctor suggested Bunker should develop "alternative methods for disposal of SO_2." Since, in his view, granting an extension would only encourage the company to continue pursuing its misguided sulfur dioxide abatement strategy, Dahlberg urged denial, "to provide added incentive to change this [company] policy of directing SO_2 only to sulfuric acid."[73]

O. B. Scott, a surgeon and physician in Kellogg since 1952, framed the issue in a commonsense way, stating, "I think it has been obvious to many people in the community when your eyes are red and they run, and you cough and gag, that there is something operating besides fresh air." While Scott was willing to grant Bunker "a short, temporary delay" in meeting the state's sulfur dioxide standards, he stressed "that sometime in the very near future we should be able to anticipate a full compliance with the law."[74]

Perhaps the most compelling testimony, however, came from Clint Waltham, Kellogg High School's student body president. Conveying what he referred to as "a direct representation of the consensus of opinion of the 600-odd young adults who make up the student body of Kellogg High School," Waltham urged the commission not to grant an extension to Bunker, stating that "to continue polluting our atmosphere at the present rate would only be adding to an already dangerous situation."[75] Waltham cited instances in which school activities had been "seriously hampered" by the sulfur dioxide situation—physical education classes were cut short and other classes disrupted by "high concentration of pollution within the building itself" on January 26 of that year, football practices the previous fall had been shortened, and visiting athletic teams had "complained of the difficulty in breathing" during sporting events held in Kellogg.[76]

Noting examples of mass deaths from air pollution crises such as the Donora, Pennsylvania, incident of 1948 in which twenty lives were lost, the student body president asked pointedly, "Is a catastrophe of this nature needed to awaken our community to the severe problem that we are facing?"[77] Though he expressed appreciation for the pollution-control efforts that had been made by Bunker Hill and awareness of the importance of economic considerations to the community, Waltham clearly put himself and the students he represented on the side of a cleaner local environment, stating "we contend strongly that economy should give way to ecology and that pollution should be stopped at any cost."[78] Questioned by a commission member who wanted to know just how much he would be willing to give up for a clean local environment—"Does this mean you would be willing to have the plants shut down? And for what period of time? And what about the loss to your parents and so on?"—Waltham stuck to his guns, saying, "Well, shutting down the plant completely is not what we advocate, but we think that the pollution should be stopped."[79]

Others spoke to specific health problems attributable to the sulfur dioxide problem. Willie Dahl discussed his fifteen-year-old son, who "has been excused

FIGURE 4.2. Bunker Hill Company Lead Smelter (foreground) and zinc plant (background) and denuded landscape.

from school by two different doctors that said he was not to enter out into this SO_2 emission." Lifelong Kellogg resident Laura Williamson summarized the air situation pithily, commenting, "The air is bad on many days, but worse on others," adding, "I was born and raised here and feel that in the last months the air pollution has been worse than ever." Williamson also related that on fall days she liked to hang clothes out to dry, but "several days I came indoors after 15–20 minutes actually gasping for breath!"[80]

Notably, none of those who testified before the Air Pollution Control Committee attempted to defend the quality of Kellogg's air. Even many who argued strongly in favor of granting the extension conceded that they did not like the sulfur-dioxide-laden air. Their argument was based largely upon concerns for the local economy and the belief that Gulf-Bunker, given more time, would find a way solve its sulfuric acid marketing problems. Those opposing the extension prioritized a more rapid cleanup of what they perceived as a badly fouled, unhealthy environment, and were skeptical that granting the company more time would bring about the long-promised cleanup of valley air. The debate over the extension divided the Kellogg community into two camps. The proextension faction's primary concern for the continued economic sustenance of the area persuaded it to permit Gulf-Bunker to continue in essentially a "business as usual" fashion, while the antiextension camp contended that the time had come for the company to take the necessary steps to improve air quality, even at the risk of impairing the local economy. While the proextension faction's arguments conform with the attitudes characteristic of *classical industrial society*, in which, to quote risk theorist Ulrich Beck, "the 'logic' of wealth production dominates the 'logic' of risk production," the antiextension faction's arguments fit well with a succeeding stage of industrial development, *advanced modernity*, wherein "the 'logic' is reversed. . . . The gain in power from techno-economic 'progress' is being increasingly overshadowed by the production of risks."[81] As this and numerous other contentious debates from the 1970s onward that pitted economy against environment would prove, the question of which "stage" of industrial development had been reached in the Coeur d'Alenes was far from resolved. The working out of these social and political issues in the mining district would prove anything but simple or easy.

In April 1972, the State of Idaho promulgated a schedule of compliance and an implementation plan that specifically addressed Bunker Hill's sulfur dioxide problem. The new rules required the company to immediately cut its level of total particulate emission to half the level recorded for 1971 and also to achieve at least an aggregate 74 percent sulfur recovery rate at its lead and zinc plants.[82] The order noted that over the past calendar year the company had averaged a sulfur dioxide level of 0.09 ppm in Kellogg, or three times the state's primary air quality standard. Henceforth, Bunker would have to achieve a maximum annual mean of 0.03 ppm and not exceed a daily maximum 0.14 ppm. The state also announced that by June 1975 the company would be required to achieve a sulfur recovery rate of 85 percent.[83] In order to better ensure compliance with the new sulfur standards, Idaho also issued a detailed list of technological control measures that the company was to implement by specific dates—for instance, it was to install a wet scrubber in the lead smelter's fuming furnace granulator and upgrade the existing wet collector in the pelletizing dryer by January 1, 1975.[84]

Notwithstanding the difficulty of complying with the state's new air standards, Gulf-Bunker officials generally preferred to deal with the state rather than the federal EPA. Company officials viewed standards proposed by Idaho's new Department of Environmental and Community Services (DECS) as considerably more lenient, and its attitude toward company compliance with regulations far more flexible, than those of the EPA. In early 1973 Bunker Hill President Frank Woodruff wrote Gulf CEO Robert Allen and summed up the status of the new air pollution regulations by saying, "The State of Idaho has proposed regulations which Bunker Hill can meet. These provide for meeting the ambient air standards through 80% emission control and intermittent control of operations. Federal regulations on which hearings were held in Boise do require 96% control and disallow intermittent operation."[85] Though the company later would find even the 80 percent standard too onerous, its pattern of preferring to deal with state regulators over generally tougher federal officials would remain the rule.

Evidence of the company and the State of Idaho's mutually congenial relations were still very much in evidence during September and October 1973—ironically, the first weeks of the severe lead pollution outbreak that followed the smelter's baghouse fire.[86] In mid-September James Bax, director of Idaho's DECS, wrote Woodruff to share some good news and congratulate him on Bunker's achievements in air pollution control. Addressing him as "Dear Frank," Bax reported that it appeared the EPA was about to adopt his department's preferred approach to Bunker's sulfur dioxide situation—the use of "best available technology" coupled with intermittent control of output when needed—and then offered compliments to Woodruff, who recently had been made Gulf's vice-president for operations and transferred from Kellogg to the forty-seventh floor of the Tenneco Building in Houston, Texas. While recognizing that "there is yet a long way to go," Bax nonetheless felt it appropriate to "give credit where credit is due, and as a public agency we feel you and your key people who have been closely involved in making this progress over the past few years are to be commended for what has been done and to continue in their fine efforts."[87] In his reply, to "Dear Jim," the Gulf Resources executive expressed both pleasure at the recent shift by the EPA and appreciation of his company's smooth working relationship with Bax's agency, writing, "It is reassuring as a citizen to know that eventually reason has prevailed in setting standards and accepting rational solutions to pollution problems. . . . I have been appreciative of the patience and pragmatic attitude that DECS has taken toward the recognition and solution of pollution problems, and feel that this attitude has been extremely important in gaining industry cooperation."[88]

The metals-laden breezes that blew through Kellogg in the winter of 1973–1974, however, brought about at least a temporary shift in the cozy relationship that had developed between Bunker Hill and Bax's agency. By May 1974, the DECS had become alarmed and was growing impatient with the large mining and smelting company. On the heels of the hospitalization of the Thomas children that triggered the emerging realization that something was terribly wrong with the environmental health situation in and around Kellogg, Bax sent off a stern letter to Gene Baker, still serving as Bunker's environmental control director.[89] Although prompted by the nascent lead health scare, Bax's missive also took Bunker to task for its failure to meet the state sulfur dioxide and particulate standards agreed to the previous year: "Stack test data have indicated that there has been nearly a four-fold increase in particulate emissions since the Implementation Plan was adopted. . . . In order to meet these standards [for ambient air quality] by the prescribed date of July 1975, The Bunker Hill Company was placed on a Compliance Schedule designed to reduce total emissions by 50% of the 1971 levels. With just over a year remaining before July 1975, it appears we are slipping further from our goal of clean air."[90] Bax went on to submit a long, detailed list of production and pollution information that the company was to submit promptly to his department, and uncharacteristically ended the letter on a somber note, warning, "The gravity of the situation requires prompt action by both this Department and The Bunker Hill Company."[91]

Responding on behalf of the company, Baker explained that Bunker had adopted "the latest technology available" to address its sulfur problems, but that unfortunately, due primarily to "the difficult meteorological situation existing in Kellogg," their pollution-control system at that point remained unable to meet the primary air standards.[92] Citing a provision of the Clean Air Act of 1970 that permitted a two-year variance for firms that, despite having made a good faith effort to comply with the regulation nonetheless had fallen short of numerical targets, Baker referenced his recent request that such a variance be granted.

There was, it must be admitted, some merit to Baker's claim regarding "the difficult meteorological situation" existing in Kellogg, which lies in a steep-walled, narrow valley, prone to a weather pattern known as temperature inversions. Inversions occur when a mass of warm air comes to sit upon a colder, heavier layer of air. The layer of relatively cold air becomes trapped by the warmer air, often at ground level. Kellogg experiences an average of 150–200 inversions per year, the bulk of which occur in the early morning hours during the late fall and winter months. Inversions created substantial air pollution challenges for Bunker Hill. The weather pattern routinely served to trap smelter and zinc plant smoke, replete with heavy metals and sulfur dioxide, at ground level, where it came into contact with people, animals, and plants. Company officials commonly complained that they would have little difficulty complying even with the EPA's sulfur dioxide standards if only their works were located on an open plain. In Kellogg, however, where inversions lasting from a single morning to a full week were the rule from November through March, the company clearly required further efforts to solve its sulfur dioxide problem.

For much of the 1970s Bunker attempted to meet Idaho's sulfur dioxide benchmarks through the use of the "best available technology"—its three sulfuric acid plants and their related sulfur dioxide control systems. But the company generally failed to comply even with these relatively lenient standards, as temperature inversions commonly raised Bunker's sulfur dioxide levels above permissible levels. The company thus pressed for the option of employing a supplemental control strategy

(SCS) during periods of temperature inversion. Implementing an SCS would involve first the use of meteorological data to predict the imminent onset of an inversion, and then the curtailment or outright shutdown of sulfur-producing systems.[93]

While the State of Idaho generally was supportive of Bunker Hill's plans to employ a combination of "best available technology" and an SCS to address its sulfur issues, the EPA vacillated in its response, sometimes approving this approach and sometimes demanding that the company reach benchmarks for sulfur capture without resort to an SCS. Because the state's standards were easier and less costly to comply with, whenever possible Bunker focused its efforts on meeting those standards and largely ignored those of the EPA. Complying with Idaho's sulfur dioxide standards and implementing an SCS, however, required the company to develop the ability to rapidly predict the onset of temperature inversions and then curtail or shut down the smelter's sintering machine. In the early 1970s Bunker began investing in personnel and technology aimed at making the Kellogg's area's weather patterns and ambient levels of sulfur dioxide far more legible. By 1976 the company was operating a weather- and sulfur-monitoring system in the Kellogg area that included five weather stations, five sulfur dioxide monitors, two meteorologists, and a weather center with an advanced computer.[94] Monitoring stations telemetered wind speed, wind direction, and ambient sulfur dioxide level data continuously to the weather center. There, meteorologists employing a powerful computer for the era predicted the imminent likelihood of a temperature inversion. Should one appear likely to occur within the next few hours, the meteorologists would advise personnel at the smelter and zinc plant to begin sulfur dioxide curtailment procedures.

"Legibility" and the New Environmental Regime

In his book *Seeing Like a State,* political scientist James C. Scott makes good use of the concept of legibility. For Scott, *legibility* refers to the ability of the leaders of large institutions such as nation-states and big corporations to clearly "see" subjects of interest to them such as citizens and customers. Per Scott, a principal reason why premodern states were so crude and often self-defeating in their efforts at administration is that they possessed rudimentary knowledge about their territories and subjects. By contrast, a hallmark of the modern state and other large, bureaucratic institutions is their possession of great amounts of information—often in the form of putatively objective facts and records—about their subjects and areas of operation and administration. Referring to the production of information such as censuses, maps, land records, mortality and unemployment records, disease rates, and productivity numbers as "state simplifications," Scott maintains that they are an integral part of a larger ambition that he refers to as an ongoing "project of legibility." He asserts that institutional leaders quite literally use their instruments of state simplification as the lenses through which they "see" and interpret the landscape and people they oversee. Without such tools, Scott argues, large-scale administrative interventions would be impossible.[95]

In Scott's view, the tool of legibility often has been placed in the service of deeply authoritarian schemes of state simplification that have undermined local autonomy and effective traditional practices. Examples include Baron Haussmann's mid-nineteenth-century reconfiguration of Paris that permanently displaced thousands of poor Parisians and made it far easier for the military to suppress popular insurrections, the brutal and ineffective Soviet collectivization of agriculture in the 1920s–1930s, and, most chillingly, the Nazis' use of Dutch census maps of Amsterdam to round up and send away the city's Jews. Although Scott's analysis and examples highlight the capacity of legibility to be employed in the service of socially destructive ends, in the case of the Coeur d'Alenes' airshed, legibility represented a worthwhile goal in the service of air pollution control. Once largely achieved, legibility acted as a potent weapon in the battle against air pollution. Even as he describes the horrific uses to which the Nazis put the powerful tool of legibility in Amsterdam, Scott recognizes the potential of using that tool for good, writing, "That legibility, I should emphasize, merely amplifies the capacity of the state for discriminating interventions—a capacity that in principle could as easily have been used to feed the Jews as to deport them."[96]

Federal and state legislation aimed at achieving major social goals would seem to fit in well with Scott's observations. Ambitious national environmental legislation such as the Clean Air Act of 1970 required reams of scientific data to determine what constituted appropriate standards of "clean air," and prodigious monitoring to ensure that new air pollution standards were being met all across the nation. The abstract concept of air quality had to be made concrete and legible for dozens of different substances and in scores of places. As a company suddenly faced with the challenge of complying with new, challenging state and federal air quality standards in the 1970s, Bunker officials had to scramble—first to understand the marks they needed to hit in order to be deemed "in compliance" and then to figure out how to achieve mandated targets for sulfur dioxide and other substances.

Bunker executives quickly learned that in order to meet the new standards (particularly employing a technologically demanding system such as an SCS), they needed to transform such formerly amorphous concepts as "Kellogg area weather patterns" and "sulfur dioxide concentrations" into highly legible, discrete entities. This led to the rapid deployment of numerous tools of radical simplification such as weather stations, computers, and sulfur dioxide monitors—all designed for the alchemy of transforming abstract phenomena into easy-to-understand numerical and verbal data. Bunker Hill's project of achieving complete legibility for the Silver Valley environment, which ultimately proved unrealizable, only intensified with the outbreak of the lead health epidemic in 1973–1974. In its wake, federal and state agencies descended upon the valley's denuded hills, seeking to confer legibility upon subjects such as the health of the community, the effect of lead upon children, and the imprint of heavy metals upon the landscape. With Bunker Hill's collapse in 1982 and the declaration of a twenty-one-square-mile Superfund site centered upon the lead smelter the following year, the EPA began a cleanup process that continues to this day. To an even greater extent than was evidenced in earlier efforts to reduce ambient sulfur dioxide levels or address the lead health epidemic, the Superfund project has levied Herculean demands for the realization of the elusive goal of complete legibility for the environment of the Coeur d'Alenes.

Regulators and Tall Stacks

In late 1973 the EPA agreed to go along with the state's sulfur dioxide plan for Bunker Hill, but not long thereafter the federal agency changed its stance and demanded tougher standards. At the time, Bunker was operating under Idaho's Regulation S, a sliding scale that permitted sulfur dioxide levels in the Kellogg area to fluctuate between 0.14 and 0.28 ppm, depending upon levels of production.[97] In November 1975 the EPA disallowed Idaho's 72 percent standard for sulfur capture and insisted that Bunker meet a rate of 82 percent.[98] The company responded that meeting the increased standard was simply unfeasible using current technology and filed a lawsuit against the EPA in the Ninth Circuit Court of Appeals in San Francisco. Bunker's lawsuit alleged that the agency had acted "in an arbitrary and capricious manner" in promulgating its new sulfur dioxide regulations.[99] The State of Idaho submitted a brief backing the company's position.

In November 1975 Gulf-Bunker, after consulting with the meteorological firm North American Weather Consultants, made the fateful decision to construct "tall stacks" at its lead smelter and zinc refinery.[100] Seeking a technological fix to its ongoing sulfur dioxide issues with the EPA, the company decided to build smokestacks tall enough to vent smelter and refinery gases at a distance from Kellogg and Smelterville and their nettlesome air-monitoring systems. The EPA opposed Bunker's decision to build the tall stacks, contending that the stacks offered no guarantee of compliance with federal sulfur dioxide and other air quality standards, and that merely dispersing the pollution over a wider area, as the stacks would do, offered no real improvement for the environment of north Idaho. EPA regional administrator Douglas Costle expressed concern that "the tall stacks will spread the harmful pollutants to outlying communities, some of which presently have much lower ambient sulfur dioxide concentrations."[101]

Idaho governor Cecil Andrus, on the other hand, expressed appreciation for Bunker's decision to build the $11.4 million stacks and hopes that the project would achieve the desired results. Echoing Bunker Hill president James Halley's statement that the tall stacks were essential to keeping the fifty-nine-year-old smelter and its five hundred workers in business, Andrus remarked, "While I remain concerned about any degradation of the air anywhere in Idaho, we must also recognize that the lead smelter is important to the lead-zinc mining industry in Northern Idaho."[102] Construction of the 610-foot zinc refinery stack began in April 1976 while construction of the 715-foot smelter stack awaited the following year. Both were completed and in operation by August 1977. The stacks towered over the valley floor and with the aid of powerful fan systems usually sent their plumes streaming up hundreds of feet.

In a significant victory for Gulf-Bunker, the Ninth Circuit Court affirmed in July 1977 that the EPA's 82 percent capture standard had been unreasonable and was not based upon well-grounded evaluation of existing sulfur-control technologies.[103] The court sent the issue back to the EPA with instructions to develop a sounder sulfur dioxide rule for the Kellogg area. At the same time, the tall stacks seemed to be accomplishing their intended goals. Bunker reported that curtailment of operations (SCS) had been "negligible" during the stacks' initial six months of operation.[104] While the company intended to maintain its weather-monitoring system so it could employ curtailment procedures as needed, the tall stacks made the system mostly irrelevant. Monitoring data also showed that while Kellogg air still did not meet federal sulfur dioxide requirements, it had improved considerably since the company began its reduction efforts in the early 1970s. When the Lurgi sintering machine went online at the smelter in 1970, ambient sulfur dioxide levels averaged 0.17 ppm, nearly six times the federal maximum. The levels fell to less than 0.09 ppm by 1973. Three years later they dropped to roughly 0.07 ppm, and in 1977, following introduction of the tall stacks, sulfur dioxide levels fell to approximately 0.05 ppm, within striking distance of the federal air standard.[105] Bunker achieved an 80 percent sulfur capture rate by decade's end, approaching the EPA's mandated 82 percent rate.[106] In June 1979, following years of struggle, the company and the federal agency reached a sulfur settlement that did not require the company to do much more than it already was doing.[107] By the end of the 1970s, although Bunker Hill still had numerous air pollution compliance issues with the federal government, and a handful with the state of Idaho, sulfur dioxide had become one of the company's smaller worries.

Greening the Country

In the early 1970s Bunker Hill decided to try on a new hat. After decades of decimating the local greenery, the mining and smelting company decided to try its hand at making the desert bloom.[108] With its barren hills and streamsides, in the early 1970s the Silver Valley more closely resembled a moonscape than a typical north Idaho landscape. Despite the obvious challenges, in 1972 the company launched an ambitious plan to revegetate eighteen thousand acres of land running from Elizabeth Park to Pinehurst and spanning elevations from 2,200 to 4,800 feet above sea level.[109] Per the company, by 1972 sulfur dioxide levels "were reduced to the point that revegetation could proceed uninhibited by smelter emissions."[110] Bunker spent the first three years of the project largely researching what species of trees and shrubs were best suited to grow in the valley's heavily eroded, metals-laden soils. In the spring of 1975, the company began its first large-scale planting efforts under the direction of University of Idaho forestry graduate student Ed Pommerening. Crews of high school and college students, along with troops of Boy Scouts, planted forty-five thousand trees in the first year. By the end of 1977 that number grew to over 157,000 trees and shrubs.[111]

Pommerening, a young forester frequently referred to as "a modern-day Johnny Appleseed" in newspaper articles, developed a number of innovative techniques in the course of the revegetation project. Learning that miners had been growing fruit trees and vegetable gardens deep in the bowels of the Bunker Hill mine for years, Pommerening decided to use an abandoned ventilation shaft at level five of the mine as an underground tree nursery. Three thousand feet below the Earth's surface, with a constant temperature of seventy-seven degrees and 100 percent humidity, the forester found that plants grew exceptionally well, and at a fraction of the cost of building and maintaining aboveground greenhouses.[112] The oxygen and carbon content of the air was ideal, as was the lack of disease and insects. The only problem, a lack of light, was remedied with the installation of mercury vapor and sodium lamps.[113] In his unique underground nursery, Pommerening grew ponderosa, lodgepole, and Austrian pines. In another innovation, the forester began growing the seedlings in soil-filled containers. The containers provided a nurturing home for the seedlings after their transplant to the hostile conditions outside the mine. As it turned out, the seedlings needed all the help they could get. The revegetation project faced severe challenges due to the relative lack of soil on the valley's steep, eroded hillsides and the high metallic content of those soils.[114] Pommerening, his curiosity and sense of personal challenge piqued by the valley's adverse conditions, commented, "I wanted to find out if something would grow on this hostile terrain."[115]

Bunker's turn at environmental remediation included not only tree planting on the hillsides but also the planting of willow shoots along the barren shores of the South Fork of the Coeur d'Alene River, shrubs and grasses on dusty former tailings ponds, and the placement of nearly one hundred thousand dollars' worth of clean topsoil on the yards of interested residents of Wardner, Kellogg, and Smelterville.[116] The company planted tens of thousands of willows and native shrubs.[117] When met with success, as were some of the company's revegetation efforts, they served both to reintroduce plant life into the local ecology and to stabilize area soils. These projects, perhaps inspired by the "environmental turn" of the 1970s, represented a worthy albeit relatively small repayment of Bunker's outsized environmental debt in the Coeur d'Alenes.

In what must have been a welcome shift from the barrage of environmental criticism and regulation that beset the company in the 1970s, Bunker was recognized for its environmental reclamation work. In mid-1975, representatives of Washington State University and the University of Idaho's Department of Forestry observed the company's revegetation efforts, prompting Halley to brag to Woodruff, "We are rapidly gaining a favorable reputation in the revegetation area."[118] The coup de grâce, though, was garnering the Idaho Tree Farmer Association's Industrial Forester of the Year award for the company's novel underground greenhouse work.[119]

While Bunker undoubtedly made some strides, the job of revegetating the Kellogg area was far from complete when the EPA assumed its direction in the 1980s. Like the company, the EPA has expended considerable resources on revegetation efforts. In an operation that the EPA farmed out to the Army Corps of Engineers in 1998, the corps deployed Sikorsky helicopters to bomb one thousand acres of hillsides south of Kellogg with grass seed. The corps turned to grass seed after experiencing very slow and stunted growth with the 150,000 white pines and various types of bushes it had planted. The federal effort, employing a sticky substance known as tackifer to keep seeds from blowing or washing away, and thousands of tons of lime to make the soil less acidic, took years and millions of dollars to complete.[120] A visit to the area in 2010, however, revealed that while the hills around Kellogg host numerous trees and grasses, the vegetation lacks the density found in neighboring forests. Rather than a thick coating of trees and grasses, there is a rather patchy forest replete with numerous bald spots, a condition that nonetheless represents a significant improvement over the nearly barren landscape of the early 1970s.[121]

CHAPTER 5

RAINING DOWN POISON

> There are always losers and winners in risk definitions. The space between them varies in relation to different issues and power differentials. Modernization risks from the winners' point of view are *big business.*
>
> **Ulrich Beck**

> If we look behind the sober scenes that the members of our bourgeoisie create, and see the way they really work and act, we see that these solid citizens would tear down the world if it paid.
>
> **Marshall Berman**

On September 3, 1973, a fire broke out in sections six and seven of the baghouse at the Bunker Hill smelter. The baghouse functioned as the principal air-pollution-control system for the company's massive smelter. Thirty-foot-by-eighteen-inch Orlon bags served to filter out many of the toxic metals and chemicals from the smelter's nearly constant stream of waste gases. The fire destroyed 2,800 of the baghouse's 12,000 bags.[1] Unfortunately, because the fire extensively damaged the smelter's roof, there was no structure to which new bags could be attached. With the smelter's pollution-control system rendered ineffectual, Bunker Hill was faced with a major decision.

In the most safety-conscious scenario, the company might have chosen to suspend smelter operations until the damaged sections of the baghouse could be repaired, though this undoubtedly would have entailed production and profit reductions.[2] Less cautiously, company officials could have chosen to significantly curtail smelter output until the damaged baghouse sections were fixed—since roughly one-quarter of the pollution control system had been taken offline, it might have seemed reasonable to decrease production by about 25 percent. A third option was to ignore public health and environmental concerns and maximize short-term profits by running the smelter full bore. In a decision with major ramifications for the future of their company and the welfare of the people and the environment of northern Idaho, company officials decided to pursue the latter course—they ran the smelter at full capacity despite the fact that its air-pollution-control system was not in working order.[3]

Although baghouse seven was replaced by November 3, 1973, baghouse six remained off-line until March 17 of the following year, over half a year after the fire.[4] In those six-plus months, considerably more than a ton of lead rained down on the people of Kellogg, Smelterville, and outlying areas of the district. In fact, as much as 2,500 tons—equal to eleven years' worth of the smelter's normal emissions—fell on the area in those few months.[5] That amounted to in excess of thirty tons of lead per square mile for Kellogg and twenty-five tons per square mile for Smelterville.[6] According to Ian von Lindern, a Yale-trained scientist whose doctoral dissertation focused on the air quality problems faced by Kellogg and its environs in the wake of the baghouse fire, dust samples taken from Kellogg homes near the smelter registered at up to 39,910 parts per million of lead, far above levels considered safe for humans.[7]

The rising price of lead may have persuaded Gulf-Bunker officials to continue operating the smelter following the baghouse fire. In January 1973, the price of lead stood at $286 per ton. By October 1974 it had risen to $479 per ton. In September 1973, when the decision was made to keep the smelter running full tilt, the price of lead was high and rising. In early January 1974, when company officials opted to bypass the remaining functional sections of the baghouse and spew unfiltered effluent directly out the smelter's main stack, the price of lead was even higher and was still climbing.[8] Indeed, the decision was rewarded by a substantial profit increase. Gulf's Bunker Hill division went from a net profit of $9 million in 1973 to a $26 million profit in 1974.[9]

This chapter analyzes the relationship between aerosolized lead and human health in the Kellogg area. It begins with the notorious lead health epidemic of 1973–1974 and ends with the lowering of Silver Valley children's blood lead levels to the national average in the 2000s. The chapter also covers the struggle between Gulf-Bunker Hill and federal and state regulators over permissible levels of lead and particulate emissions and appropriate pollution-control technologies. It addresses the concept of "legibility," and how both the company and environmental regulators attempted to make air pollution in the Coeur d'Alenes fully transparent and readable. Finally, the chapter addresses the shutdown of the Bunker Hill smelter in early 1982 and the effects that this event, along with twenty-five years of subsequent Superfund interventions, have had on the environmental and health landscapes of the area.

This chapter spotlights the inherent risks and environmental destructiveness of industrial modernity and demonstrates the tendency of those risks to be unfairly distributed, both economically and geographically. As Beck observes, "The history of risk distribution shows that, like wealth, risks adhere to the class pattern, only inversely: wealth accumulates at the top, risks at the bottom. To that extent, risk seems to *strengthen*, not to abolish, the class society."[10] This was certainly the case with the 1973–1974 lead health crisis, which most severely impacted those living closest to the smelter in relatively poor neighborhoods such as Deadwood Gulch and Smelter Heights, and in the town of Smelterville. In addition, the chapter points to the role of decision making in the "risk society." While social risks increase in the move from a preindustrial to an industrial society, specific individual and corporate decisions within industrial society can increase or decrease social risks. In the case of the 1973–1974 Kellogg lead epidemic, the decision of Gulf-Bunker Hill to run its smelter full bore for months following the baghouse fire caused major environmental damage and significantly heightened the risk and social costs of living in the Coeur d'Alenes, particularly for children living within a few miles of the smelter. It also must be noted that for some places in the United States the "price of progress" has been far steeper than for others. This is an unhappy truth—the health and environmental costs of industrial modernity have been shared unequally. Some communities that have contributed significantly to the nation's industrial progress, Kellogg and Toms River, New Jersey, for example, have suffered mighty health and environmental pains for their efforts, while other communities that have contributed far less have become much wealthier while remaining far less polluted. Modernity's tendencies toward "Faustian development," to use Berman's phrase, are on ample display in the chapter.[11]

Gulf-Bunker officials were well aware of the health risks associated with high levels of airborne lead. In fact, they had become somewhat preoccupied with a recent outbreak of lead intoxication in children living near ASARCO's lead smelter in El Paso, Texas. In a pathbreaking 1970 study that raised national attention, Dr. Philip J. Landrigan of the federal Centers for Disease Control (CDC) found that children residing nearby the ASARCO smelter suffered from abnormally high, and potentially dangerous, levels of lead in their bloodstreams. In 1972 the City of El Paso sued ASARCO, and, following several weeks of testimony, the mining and smelting giant agreed to settle out of court.

Leaded Kids and the Company

Alarmed by the potential implications of the ASARCO/El Paso situation for their operation, Gulf-Bunker began quietly investigating the details. After Idaho's Department of Health and Welfare notified Gulf officials that higher levels of lead had been measured in Kellogg than were found near ASARCO's El Paso smelter, the company also began quietly sampling soil near the smelter and hired a young Kellogg physician, Ronald Panke, to begin testing local children for lead intoxication.[12] The soil samplings turned up evidence of very high levels of lead and cadmium, with lead measuring over 5,000 ppm.[13] In his study in 1972, Panke performed blood and urine tests on ninety-nine children living near the smelter.[14] The doctor's tests showed the presence of elevated lead levels in 34.4 percent of the children.[15] Despite this early evidence of apparent lead health issues in a significant number of children living close to the smelter, the company did nothing to alert parents of their children's high lead burdens, nor did it alter its smelting practices. Neither did Gulf-Bunker warn residents adjacent to the smelter of the likelihood of high concentrations of dangerous metals in their yard soils. In addition, Panke's request to Gulf-Bunker to perform follow-up studies to verify his findings went unheeded.[16] The company, evidently satisfied with possessing greater knowledge of the lead health situation in Kellogg and Smelterville, felt no need at the time to act on that information.

Their study of the El Paso lead intoxication experience, however, does appear to have influenced the Gulf-Bunker chieftains' decision-making process in the period immediately following the baghouse fire. Internal company notes handwritten by Gulf vice-president and former Bunker president Frank Woodruff indicate the presence of straightforward cost-benefit calculations in Gulf-Bunker's approach at the time. Woodruff wrote, "El Paso—200 children—$5 to 10,000 per kid."[17] An unidentified Bunker Hill official responded to Woodruff's note with an estimate that Gulf's legal liability for poisoning five hundred Kellogg children would be "$6–7 million."[18] Based upon the company's profit increases in 1974 over the previous year, and accepting the company official's estimate of Gulf-Bunker's legal liabilities for poisoning Kellogg children, the company would have come out $10–11 million ahead if forced to pay this amount of compensation. Additional handwritten memos that came to light following the unsealing in 1990 of records from the Yoss case against the company on behalf of nine lead-intoxicated Kellogg children, show that Gulf executives examined other alternatives: cleaning up the smelter constituted one option, discrediting doctors who warned of the dangers of lead poisoning represented another.[19] Gulf-Bunker's apparently calculating, bottom-line approach to its airborne lead problems, while financially beneficial to the company in the short run, ultimately led to a bevy of plaguing problems for both the company and the people of the Coeur d'Alenes.

In early April 1974, shortly after the company finally fixed its baghouse problem, a pair of young schoolchildren visited a Kellogg doctor complaining of flulike symptoms. But they did not have the flu. Hospitalized for severe lead intoxication,

FIGURE 5.1. Kellogg area children awaiting blood lead screenings, August 1974.

the Thomas children's illness triggered a major lead scare in the Coeur d'Alenes. Represented by their parents, the children sued Gulf for $500,000 in general damages and $500,000 in punitive damages.[20] In early May, Idaho DECS field investigator Ian von Lindern reported to Idaho Panhandle Health District official Jerry Cobb that ambient air lead concentrations in Kellogg were running at 16.1 micrograms per cubic meter (μg/m^3) of air, nearly double the rate in 1972 and 1.5 to 5 times the concentrations found in El Paso.[21] Von Lindern's soil samples showed concentrations of lead that were about twice as high as those found in El Paso.[22] In late May, James Bax of the Idaho Department of Health and Welfare wrote Bunker's Gene Baker, expressing concern over environmental quality in the Kellogg area and requesting data on production and sulfur dioxide emission rates.[23] A week later, meeting in Boise, personnel from the federal Department of Health and Welfare and the CDC agreed to perform environmental and blood sampling during the summer in the Kellogg area.[24] In August, in a study overseen by the CDC, blood draws and interviews were performed on 919 children in five study areas of varying radii from the smelter stack and 137 children in control areas removed from air currents affected by the smelter's waste gases.[25]

The study's findings, released in September, were shocking. Within one mile of the smelter, 170 of 172 children, or 98.8 percent, were found to have at least an "elevated" blood lead level of 40 μg/dl.[26] Nearly a quarter of the children, 38

TABLE 5.1. Mean blood lead levels of children at varying distances from the smelter stack in 1974 and 1975 (μg/dl)

	1 mi. or less	1–2.5 mi.	2.5–6 mi.	6–15 mi.
August 1974	68.3	49.1	34.7	33.3
August 1975	47.1	39.3	31.4	26.7

Source: Glen Wegner, *The Shoshone Lead Health Project: Work Summary* (Boise, ID: Idaho Department of Health and Welfare, 1976), 78.

(22.1 percent), had levels of 80 μg/dl or higher, far above the 60 μg/dl threshold for emergency hospital treatment.[27] The median blood lead level for these children living nearest the smelter was 68.3 μg/dl.[28] One-year-old Arlene Yoss suffered from the highest concentration of lead ever recorded in a human's bloodstream, 164 μg/dl.[29] All of the parents of children with blood lead levels of 80 μg/dl or higher were informed in September that they should place their children under medical care. Arlene Yoss was hospitalized in Coeur d'Alene on September 9.[30]

Not surprisingly, blood lead concentrations decreased with increased distance from the smelter. Between one and two and a half miles from the smelter, the number of children with "elevated" blood lead dropped to 151 out of 199, or 75.9 percent, and the number requiring immediate hospitalization decreased to three. A high correlation was found between elevated blood lead levels and high concentrations of ambient lead in the air. Between October 1973 and March 1974—when the lion's share of the emissions occurred—the amount of ambient lead in the Kellogg air increased by 50 percent, from an average of 6.7 $\mu g/m^3$ over the previous year, to 10 $\mu g/m^3$.[31] In January 1974, acquiescing to the request of overwrought baghouse workers, the company agreed to bypass the baghouse and vent unfiltered smoke and gas directly into the atmosphere. An average of sixty tons of lead a month—four times the average for the previous six years—spewed from the stack through March.[32] After entering the airstream, and from there the lungs and bloodstreams of the people of the Coeur d'Alenes, the lead particulate matter eventually fell to the ground and settled into the dirt and plant life, to again be ingested by people and animals.

Lead and Horses

On April 25, 1974, Baker mailed a letter to landowners in the Kellogg area, alerting them to the risk involved in pasturing livestock, and particularly horses, on local lands where smelter toxins tended to fall in heavy quantities.[33] For years Bunker Hill had been compensating area horse owners for what the company deemed le-

TABLE 5.2. Mean ambient air lead concentrations at varying distances from the smelter stack, 1974

1 mi. or less	1–2.5 mi.	2.5–6 mi.	6–15 mi.
18 μg/m³	14 μg/m³	7 μg/m³	3 μg/m³

Source: Wegner, *Shoshone Lead Health Project*, 72.

gitimate claims of loss. Indeed, since 1920, just a few years after the smelter began operating, the company had been buying smoke easements, and making direct land purchases, from area landowners.

In the 1970s, Gulf-Bunker employed a relatively routine procedure for dealing with horse deaths in the Kellogg area. If Roy Larson, a local veterinarian, determined that an animal had been sickened or killed by heavy metals such as lead or cadmium, the company would make a payment for the value of the lost horse. If, however, Gulf-Bunker believed the horse death to have occurred outside of the region that its smelter could have polluted, it would deny claims, as it did in the 1974 case of two deceased horses that had been pastured near the Rio Vista Terrace.[34] Larson's reports included the cause and the place of death, noting, for example, that the horse had suffered from symptoms that "were typical for lead and antimony poisoning and the horse was in a smelter smoke area near Kingston."[35] Sometimes the payouts were relatively large, as in a settlement in 1973 involving the deaths of four registered Morgan horses that recently had been transplanted to the area. That sad case cost the company $2,842.00 (or $15,700 in 2011 dollars). Sometimes Gulf-Bunker would warn a landowner who had been paid repeatedly for horse losses on a particular pasturage that the land was clearly toxic to the animals, and that the company would no longer offer compensation to individuals foolish enough to place their animals on those locations.[36]

A Lead Health Study Is Born

Gulf-Bunker officials knew they needed to do something in response to the outbreak of lead poisoning adjacent to its smelter. After the Idaho Department of Health and the CDC released their findings in September, the national media ran with the story of Kellogg's "leaded children," shining unwelcome publicity on the company. In October company officials decided upon a course of action—they would sponsor a major medical study of lead pollution, children's health, and the environment in Shoshone County. A company-funded, industry study, they believed, would address the public's fears about lead contamination in and around Kellogg and would have the added advantage of likely preempting a potentially hostile CDC study. Gulf-Bunker first attempted to partner with the International Lead Zinc Research Organization (ILZRO), an arm of the industry's trade organization that had conducted research following the lead scare in El Paso. That effort, however, was quickly derailed by loud complaints from within the state government that such an industry-financed, industry-run study would not pass muster with the public. The study was reinaugurated, however, later in the month, as the Shoshone Lead Health Project. The company now would work with Idaho's Department of Health and Welfare as a joint partner, with the company footing 95 percent of the bill. Governor Andrus appointed Glen Wegner, a practicing physician and attorney from Boise, to be the project's director.[37] The project was to be overseen by a technical committee that included a number of senior figures from the fields of pediatrics and toxicology, including J. Julian Chisolm Jr., a leading expert on pediatric lead intoxication from Johns Hopkins Hospital.

After fifteen months of work, the Shoshone Lead Health Project released its final report in January 1976. The report's authors concluded that "at the present time and in our best judgments, we do not feel any permanent clinical impairment or illness [to the children] has occurred. Further, it is not likely to occur in the future due to this particular exposure."[38] The technical committee noted, however, the presence of "mild anemia" and "minimal degrees of slowing in nerve conduction" in those children who had experienced the highest exposures to lead, although it concluded that "there is no evidence to date that these effects will be in any way permanent."[39] The report also called for the continued monitoring through 1980 of the lead health status of Kellogg-area children, both those who had lived through the 1973–1974 outbreak and those who moved into the area over the next few years. Finally, the technical committee urged "all responsible parties—industry and government in particular—to continue to do their utmost to avoid a replication of the precipitating events."[40]

Reading beyond the Shoshone Project's final conclusions, however, reveals a more complex and less reassuring portrait of the Kellogg area's environmental and health situation in the mid-1970s. The report's environmental sections, authored by A. F. Yankel and Ian von Lindern of the Idaho Department of Health and Welfare's Division of the Environment, leave little doubt regarding the cause or magnitude of the air lead and human health problems in the Kellogg area. Noting that "levels of lead higher than those found in the existing literature are found in the Kellogg-Smelterville area," Yankel and von Lindern concluded that the Bunker Hill smelting complex was the chief cause of the abnormally high levels of lead found in the area's air and soils.[41] Acknowledging the presence of other active sources of lead—old mining and milling centers, floodplains where tailings had accumulated, ore trains that spilled some of their contents—von Lindern and Yankel nonetheless determined that "generally, it is apparent that the center of abnormal lead concentrations in the valley is the Bunker Hill complex . . . in both magnitude and source strength, the largest and certainly what is expected to be the most significant active source of contamination is air-borne emissions from the Bunker Hill complex."[42] Yankel and von Lindern's studies revealed that the dilution pattern of heavy metals in the area was logarithmic with respect to the smelter, with concentrations falling as distance from the smelter increased.[43] The highest statistical correlation for children's blood lead levels was found with levels of lead in the air (0.72), but blood lead levels also correlated strongly with the levels of lead found in the soil (0.55) and house dust (0.24)—all were higher the closer one got to the smelter.[44]

The environmental sections of the project's final report evoked a landscape coated in lead and other heavy metals (relatively high levels of cadmium, arsenic, antimony, and mercury also were found in areas where lead particulate was found in quantity). Not surprisingly, the metallic coating was thickest in Smelterville and Kellogg, the communities closest to the lead smelter. Testing done in September 1974, a year after the baghouse fire, showed levels of lead in blood, air, soil, household dust, and on homegrown vegetables in Smelterville and Kellogg that were off the charts—lead levels this high had never before been measured in North America.[45]

Although the correlation between levels of air and blood lead was the strongest found, the environmental scientists and doctors did not believe that simply breathing the heavily leaded air constituted the primary means by which children attained high levels of blood lead. Of greater significance was the potent dust that was formed when the heavily leaded air mixed with soils in the area near the smelter.[46] The dust found its way into homes in neighborhoods such as Deadwood Gulch and Smelter Heights, where it was easily inhaled or otherwise ingested by children. The formation of house dust made lead and other heavy metals more readily bioavailable to Silver Valley children.

While Yankel and von Lindern learned that the most important factor in determining excess lead absorption in area children was proximity to the smelter, other factors also were found to be significant. Children's ages and the hygienic condition of their homes both affected their levels of lead absorption. Children between the ages of two and four were most susceptible to high levels of lead absorption, likely because of their tendency to play in the dirt and put objects in their mouths. Likewise, homes containing the highest levels of dirt and dust tended to house children with the highest blood lead levels.[47]

The environmental report also showed that, because of the pollution burst that followed the baghouse fire, some area children had received extremely high levels of lead exposure at the most susceptible ages for development. Yankel and von Lindern concluded that children between one and three years of age in September 1975 had, through their lifetimes, "experienced external exposures 1.5 to 2 times greater than the older children in the community."[48] Drawing upon evidence offered by the high-volume air samplers that had been monitoring the quality of Silver Valley air at different locations since 1971,[49] the scientists determined that a Kellogg child who was born in 1973 would have experienced a lifetime average air lead exposure of 17.5 μg/m^3, while for a Kellogg child born in 1966 the average air lead would have been 11 μg/m^3. The child born earlier also would have experienced a far lower level of exposure during the critical first three years of life.[50] For children growing up in neighborhoods within half a mile of the smelter stack, air lead exposures would have been approximately even 50 to 75 percent higher than those found in Kellogg.[51] Summing up the air pollution situation in Kellogg and its environs two years after the baghouse fire, von Lindern and Yankel said, "The exposure problem in Kellogg can best be described as an acute period of extremely high air lead exposures, lasting about two years, imposed on a chronic lower level of abnormal exposure from several sources."[52]

Yankel and von Lindern concluded by calling for the continued monitoring of the air conditions in the Silver Valley and expressing concerns that despite progress in air quality, significant strides still were needed if the goal of bringing the valley's population below the clinically "elevated" blood lead level of 40 μg/dl by 1977 were to be met.[53] While air and blood lead levels both dropped significantly from September 1974 through September 1975, with mean blood lead within a one-mile radius of the smelter falling from 68.3 μg/dl to 47.1 μg/dl, there was reason to believe that this pace of improvement in the area's worst-affected region was not sustainable. For one thing, the numbers from 1975 were skewed by the fact that so many of the children tested the year before had moved out of the area when the homes in Smelter Heights and Deadwood Gulch were purchased and demolished by Bunker Hill as a part of the company's new legal-environmental strategy. Of the forty-two children tested in 1974, only seven were retested the following year. In light of evidence that the project's blood lead level goals were unachievable within the proposed time frame "without taking extreme measures," von Lindern and Yankel questioned whether the goal of achieving 40 μg/dl should be modified.[54]

Philip J. Landrigan, the Harvard-trained medical doctor who a few years earlier had led the investigation of the El Paso lead health episode, spearheaded the medical component of the study. Landrigan, who now served as chief of Environmental Hazards Activity, Cancer and Birth Defects Division, Bureau of Epidemiology, at the federal CDC, led a nine-member team of medical doctors and scientists who authored a section of the project's report titled "Increased Lead Absorption with Anemia and Slowed Nerve Conduction Velocity in Children Near a Lead Smelter." His coauthors included Edward L. Baker, a medical epidemiologist for the CDC; John A. Mather, MD, Idaho's state epidemiologist; and Yankel and von Lindern.

The Landrigan group noted the unprecedented severity of the Kellogg situation, commenting, "To our knowledge, this is the highest prevalence of increased lead uptake ever recorded among children in a community."[55] The team also concluded that the smelter was the principal cause of the lead intoxication crisis and that airborne lead oxide along with lead oxide and lead sulfide in household dusts and outside dirt formed the chief means by which lead entered the bodies of Silver Valley children.[56] Turning to diagnosis, the team found that while no clinical conditions or overt neurological toxicity had surfaced among Silver Valley children, various forms of lead-induced impairment were in evidence. They also reported that in addition to such environmental factors as proximity to the smelter and levels of household dust, social factors such as status, father's employment, and length of residence in the area played a role in determining blood lead levels. In all but one of the five geographic areas used in the study, the highest blood lead levels were found in children from lower social strata.[57]

The medical team found that the children of smelter workers generally had considerably higher blood lead levels than did the children of miners, and that in turn, the children of miners tended to have higher blood lead levels than did the children of nonminers living in the same areas.[58] This suggested that smelter workers (and to some extent miners) who brought home dust on their clothing and skin contributed, however unintentionally, to the "leading" of their children. Length of stay in the area also correlated with higher blood lead levels. The mean blood lead level of children who had moved into Area 1 within six months of the study was 55.9 μg/dl, significantly less than the overall mean of 68.1 μg/dl for the area. In addition, free ethrocite proporphyryn (FEP) levels, another measure of lead burdens in humans, tended to rise progressively during the first four to six months of a child's residence in the area.[59] All of this suggested that, excluding children who ate paint or other leaded materials, and recognizing that certain social factors tended to lead to higher levels of blood lead among children (e.g., low social status or having a father who worked at the smelter), simply living in the Coeur d'Alenes was enough to produce elevated blood lead levels for many of the area's children.[60]

Based upon earlier findings showing that occupational lead exposure had led to slowed nerve conduction velocities among workers, the Landrigan team decided to test nerve conduction velocities among Silver Valley children.[61] Using electrodes to provide stimulus, Landrigan tested nerve conduction velocity in 202 apparently healthy Shoshone County children. While "no frankly pathologic conduction velocities were noted," the medical team found that nerve conduction velocity was slowed among children with higher blood leads.[62]

In addition to FEP elevations, Landrigan also found that anemia was considerably more likely to be found in children with the highest blood lead levels. While only 2.2 percent of all the children examined were anemic, 17 percent (seven of forty-one) of those with "clinical" blood lead levels of 80 μg/dl or higher were. The Landrigan team's research indicated the presence of a dose-response effect with regard to environmental lead in the children of Shoshone County. The anemia and slowed nerve conduction velocities found in more highly lead-intoxicated children, while not in themselves life-threatening conditions, nonetheless represented real physical harm done by the Bunker Hill smelter to some Kellogg area children.

The technical committee's chairman, Dr. Wegner, and two medical doctors from Idaho's Department of Health and Welfare, John A. Mather and J. T. Ashley, recommended continued monitoring of the lead health situation in Shoshone County. The doctors noted that in addition to providing valuable scientific information, continued monitoring efforts would offer early detection of any regression on the air and blood lead fronts and help prevent the outbreak of another "lead absorption epidemic."[63] The group called for annual blood lead and FEP testing of those they viewed as the most seriously affected of all Shoshone County children—one-to-eight-year-olds living in Areas 1 and 2 (within 1 and 2.5 miles of the smelter, respectively).[64] In addition, they recommended testing all one-to-eight-year-olds who moved into Areas 1 and 2, first upon their arrival and annually thereafter. The group of doctors also called for a one-year study of cord blood lead in all Shoshone County newborns, to determine how much lead area babies were absorbing in utero.[65]

In their most ambitious recommendation, Wegner and the other doctors suggested testing the approximately eighty children born in Area 1 during the levels of highest lead emissions for neurological damage. Noting that "children under one year of age are purportedly more susceptible to neurological damage from lead than older children," and that no Shoshone County children had yet been tested for such damage, the Wegner group called for neurological tests to be performed at the University of Washington Medical School (or a comparable facility) and for those results to be compared with those of a control group of children that had not been exposed to high amounts of lead.[66] The doctors called for this monitoring and testing program to last until June 1980 or until the last Shoshone County child's blood lead fell below the 40 μg/dl threshold, whichever came first.

While the Shoshone Lead Health Project did not unambiguously demonstrate that the children of Shoshone County had been irreparably harmed by the lead and other toxic emissions from the Bunker Hill smelter, it raised serious questions about the effects of the 1973–1974 lead epidemic on the health of area children. Offering considerable evidence of the effects of the smelter's waste gases and smoke on the human health and environment of the region, the project clearly indicated that the smelter was the chief culprit in the lead intoxication epidemic that followed the baghouse fire and that over decades it had played a primary role in permeating the soils, dusts, plants, and animals of the Coeur d'Alenes with elevated levels of lead and other toxic heavy metals. These toxins could be found wherever the winds had blown them after they left the smelter stack and they often remained readily bioavailable for decades or more.

Smoke Gets in Their Eyes

The 1970s, as noted, ushered in a new and exponentially more challenging era of environmental regulation for Bunker Hill. While the company's most challenging and prolonged air-related environmental battle concerned sulfur dioxide, it also was forced to respond to strict new federal standards for both particulate matter and ambient levels of lead in the air.[67] The Clean Air Act of 1970 promulgated new standards for particulate matter and sulfur dioxide. Specific compliance regimes and attainment deadlines were left up to the states to determine on a case-by-case basis.[68] Standards for airborne lead, however, were not included in the Clean Air Act. In part due to the Kellogg lead health scare of 1973–1974, some health officials, including the CDC's Philip Landrigan, began pushing for a federal airborne lead standard of 1.5 $\mu g/m^3$. Although Bunker Hill and the Lead Industries Association (LIA) vigorously opposed this standard on the grounds that it was overly burdensome and unnecessary on public health grounds, the EPA adopted the regulation in 1978.

The term *particulates*, in the EPA's parlance, refers to small bits of airborne material that darken the air. For laypeople it is better known as smoke. Though often visible in the airstream, particulates range from as small as less than 1.5 micrograms to as large as 50 microns.[69] Aerosolized lead leaving the Bunker Hill smelter's stack and traveling into the Kellogg airstream, for example, would have constituted only a portion of the total particulate matter emitted by the smelter.

What regulators referred to as *particulates* and locals as *smoke* made up the most obvious manifestation of air pollution in the Kellogg area, where the gray pall of smelter smoke and sulfur-dioxide-denuded hillsides served as constant reminders of the environmental costs of the lead smelting and zinc refining. Kellogg doctor Keith Dahlberg's letter in 1971 to the Idaho Pollution Commission reflected local concerns on the subject. Dahlberg wrote, "On almost any day . . . great pillars of black or brown smoke can be seen rising, not from the smoke stacks but from the roof of the Smelter, particularly in the area of the blast furnace,"[70] and added that "there are ventilating fans in these areas, but they do not appear to be connected to any conduit taking them to the smoke scrubber." Following over eighteen months of study and discussions, in early 1973 the State of Idaho and Bunker Hill agreed to an implementation plan for particulates. This highly detailed compliance schedule required the company to install numerous new pieces of equipment and make many improvements to existing equipment in the lead smelter and the zinc refinery. The company, for instance, agreed to make baghouse upgrades that would cut blast furnace emissions from 465,000 to 330,000 pounds per year by January 1, 1974, and to install a mist eliminator to lower particulate emissions at its No. 1 acid plant from 2,760,000 to 276,000 pounds per year by January 1, 1975.[71] The plan required the achievement of a 50 percent general reduction in particulate emissions below 1971 levels by July 1, 1975.[72]

In May 1974, Gene Baker wrote Lee W. Stokes, the assistant administrator of the Idaho Department of Environmental and Community Services (DECS), to brag and explain. In the letter, Baker touted the company's "considerable progress" in performing all of the pollution control work required under its compliance schedule, completing five of the twenty-five projects "well ahead of schedule," with sixteen "on schedule" and only four behind schedule. Explaining that the four projects were behind schedule due to unavoidable delays in procuring equipment, Baker requested extra time to complete the projects.[73] Baker received a reply not from Stokes but from an obviously perturbed James Bax, the DECS's chief, who took the company to task for "nearly a four-fold increase in particulate emissions since the Implementation Plan was adopted" in February 1973.[74] Apparently disregarding its compliance schedule, the company also dramatically increased not only its lead but also its particulate emissions during a several-month period immediately following the baghouse fire. In his letter, Bax also noted that three particulate compliance projects would be over five months late in completion and a fourth project, the return sinter tank, over a year late.[75] In addition, the state official reprimanded Bunker Hill for "the notable absence of both production data and consistent emission data," which it was supposed to report regularly.[76] Bax tasked the company with immediately catching up on its reporting requirements and supplying all the missing back data within three weeks.[77] The regulator's sudden shift to a position of acute vigilance vis-à-vis Bunker Hill highlights the capacity of a catastrophe, or the threat of one, to powerfully reorient social relations in an advanced modern society.[78]

Mr. Hooper Comes to Kellogg

Mark H. Hooper, chief of the Air Technical Compliance Section of the EPA's Region Ten Enforcement Division, visited the Bunker Hill lead smelter in 1974. He found pollution-control measures there to be far below the industry standard. Addressing the smelter's problems with particulates, Hooper noted, "Large quantities of particulates are apparently emitted from the numerous stacks in the smelter and escape from the sintering, blast furnacing, and ore preparation operations as fugitive emissions."[79] Though Bunker often maintained that it employed the "best available technology" in its efforts to control sulfur dioxide releases, Hooper's observations were to the contrary. The regulator critiqued the company's particulate emissions control efforts, writing, "I made determinations based on my knowledge of accepted methods for control of fugitive emissions from materials handling and operations that the systems which are being used by the Company were inadequate. In some cases, controls or captive systems are non-existent."[80]

Comparing the Bunker Hill smelter's pollution-control practices with those he had recently witnessed at three lead smelters in southeastern Missouri's historic "lead belt," Hooper noted that while "an acceptable level of fugitive emission control" had been achieved at those facilities, the same could not be said of the Kellogg operation.[81] The EPA official judged the pollution issues at the Missouri smelters to be "extremely small in comparison to the problems which currently exist at the Bunker Hill smelter."[82] The difference, according to Hooper, was that the Missouri smelters made use of particulate control methods and technologies that Bunker either had not put into practice or had not purchased. The company's faulty baghouses evidently were a big part of the problem. Hooper wrote that "furnace exhausts and sinter tail gas presently flow to baghouses. Those baghouses, however, are inadequate, in that large quantities of particulate are emitted from them. Well designed, operated and maintained baghouses in this application should be able to reduce these levels significantly and virtually reduce plume opacity to zero."[83] Another major source of particulate emissions was "the open transfer points and unhooded conveyor belts" at the smelter's sintering plant.[84] Hooper suggested the best solution for this problem would be "to totally enclose all outside belts and transfer points" in the system,[85] as he had seen done to good effect at the Missouri smelters.

Addressing another primary source of particulate emissions, blast furnace "upsets"—periods during which the smelter's blast furnace released exponentially greater levels of pollution than at other times—Hooper recommended making improvements to its hooding and exhaust system that were similar to what was used at the Saint Joseph, Missouri, smelter. The EPA official also was unimpressed by Bunker's use of scrubbers as a principal means of particulate control, noting, "The company appears to be incapable of running them [wet scrubbers] effectively."[86] Hooper's view of company pollution control practices did not improve with his visit to the zinc plant, where he found the particulate control system to be "entirely inadequate to handle all the fugitive emissions."[87] There again, the inspector saw the solution lying in simply installing proven technologies.[88] As Hooper's investigation makes clear, by the 1970s (if not earlier), despite its oft-cited expenditures on air pollution control technology, Bunker Hill had fallen well behind industry standards in this area.

In April 1974, the EPA issued a primary standard for particulates of 75 $\mu g/m^3$, along with a maximum permissible twenty-four-hour average of 260 $\mu g/m^3$.[89] When, on July 1, 1975, Bunker as anticipated failed to meet the terms of its compliance schedule for particulates, the company fell afoul of state and federal mandates. In September the EPA issued a compliance order requiring the company to take action to control its particulate emissions and established a new compliance schedule deadline of July 1978.[90] As it had with sulfur dioxide, Bunker Hill turned to the tall stacks as its primary means of addressing its particulates problem. The company also, however, continued its efforts at the blast furnace and attempted to increase ventilation at the sinter tunnel, the tap floor, and the feed belts.[91] In addition, at a cost of $1.5 million, Bunker built a fourteen-square-foot uptake and steel flue to capture and transfer blast furnace emissions to the baghouse. By the end of 1977, the company was capturing 90 percent of its particulate emissions.[92]

Despite the company's improvements, in 1977 the EPA sued Bunker in federal court, citing its zinc fuming furnace at the lead smelter as a source of "excessive particulate emission." The agency asked the court to fine Bunker Hill $25,000 a day until it came into compliance.[93] That December Idaho's Department of Health and Welfare proposed declaring Kellogg a "non-attainment area" for particulates and sulfur dioxide. The company charged that the EPA's mandate for the zinc fuming furnace required high costs for paltry pollution gains and vowed to fight the lawsuit.[94] Bunker also objected to the EPA's additional particulate control mandates, claiming that "the remaining emissions come from hundreds of tiny contributive sources, or from specific exhaust points where the cost of controlling the emissions is prohibitive in relation to reduction to be gained."[95] The company declared Idaho's determination to be flawed because it did not account for background levels of particulates in the area's environment.[96]

On the issue of background levels of particulate matter, Bunker Hill, it must be admitted, had a point. Background levels made up a not insignificant portion of the total suspended particulate burden in the Silver Valley. A study in 1977 commissioned by the Idaho Department of Health and Welfare (DHW) estimated a mean background particulate concentration of 35 μg/m^3 in the area stretching from Wallace to Cataldo.[97] The valley contained numerous sites—roads, railroad beds, and old settling ponds, among others—where strong winds readily could lift metals-laden dust and blow it about. In an extreme incident that took place in the fall of 1972, a blinding dust storm caused eight automobile wrecks on Interstate 90 near Kellogg.[98] In a study of particulate distribution in the Silver Valley from 1977–1978, the environmental firm PES found that the contributions of smelter emissions to the total particulate burden decreased with distance from the smelter, ranging from 74 percent at Silver King Elementary, one-third of a mile from the main stack, to 8 percent at Cataldo, ten miles distant.[99] The role of fugitive dust in the particulates problem generally assumed greater prominence farther away from the smelter, but strong winds could make dust a significant issue anywhere in the valley. While the Kellogg area fell in compliance with state and federal mandates for particulate emissions on an annual basis, it exceeded limits for daily emissions on two days. Since the emissions on those days contained little lead and were recorded at points relatively distant from the smelter, PES concluded that they likely were due to a combination of high winds and open fugitive sources.[100] As Bunker developed greater control over its particulates emissions, as it was doing by the late 1970s, the need to deal with open sources of fugitive dust became increasingly apparent.

As with sulfur dioxide, Bunker Hill's investment in the tall stacks and in systems that increased draft effectively solved the company's regulatory problems concerning particulate matter. Between the July 1974–June 1977 period and the October 1977–September 1978 period, area particulate levels decreased by 30 percent.[101] By the fall of 1978, the company was meeting the EPA's 75 μg/m^3 annual particulates standard, even at the Silver King and Smelterville air sampling stations.[102] Bunker's only issue remained the occasional daily particulates average, and the company's ability to control this was limited due to the role of fugitive dust in the problem. Although the tall stacks dispersed Bunker Hill's noxious airborne wastes away from the Kellogg area, the net effect (as the EPA had predicted) was simply to broadcast particulate matter, sulfur dioxide, and lead to what had been more pristine locations. Bunker Hill nonetheless was, by 1978, undoubtedly a good bit less dirty environmentally than it had been a few years earlier. The EPA entered discussions with the company about settlement of its lawsuit of 1977 over particulate emissions. Settlement by 1979 was a viable option, said the EPA's Jim Sweeney, "because of the things they [Bunker Hill] have done to correct problems."[103]

Bunker Hill's Heavy Lead Burden

In September 1978, the EPA promulgated the first federal rules for ambient levels of lead in the air, having adopted the ambitious goal of bringing the blood lead levels of all American children below 30 μg/dl. Since levels of air lead represented the most statistically significant factor in determining blood lead levels, the EPA ambitiously set the ambient lead-in-air standard at 1.5 $\mu g/m^3$. Bunker Hill and the US lead mining and smelting industry as a whole vigorously opposed the new standard. Bunker Hill's position was unequivocal: "The nonferrous mining and smelting industry is opposing this unrealistic standard. Testimony before a recent EPA hearing on the matter indicated such a standard could not be met in the Kellogg area, nor in the vicinity of 80 percent of the primary and secondary smelters in the United States."[104] Though the company had reduced its lead emissions significantly from the stratospheric levels recorded in 1973–1974—bringing them below pre-baghouse-fire levels—the 1.5 $\mu g/m^3$ standard remained far from its reach. In May 1971, for example, a fairly typical month for air lead levels at Smelterville City Hall that year, lead concentrations registered 4.08 $\mu g/m^3$.[105] During October 1973, Smelterville lead levels spiked to 24.81 $\mu g/m^3$, more than six times an average reading in 1971 and nearly seventeen times the standard in 1978. By 1975 lead in the Smelterville air was returning to 1971 concentrations, with an average month measuring about 4.85 $\mu g/m^3$.[106] The shocking lead concentrations of 1973–1974, which registered as high as 101.98 $\mu g/m^3$ at Silver King Elementary, seemed to be a thing of the past by the second half of the 1970s.[107]

PES found that by 1977–1978, following the erection of the tall stacks, air lead levels had decreased by 40 percent from 1974–1977 measurements.[108] The highest concentrations were found at Silver King Elementary, where during the 1974–1977 period the average concentration of air lead measured 16.2 $\mu g/m^3$. Although reduced by 1977–1978 to 9.2 $\mu g/m^3$, the air lead concentration at Silver King remained nearly seven times the EPA standard.[109] The PES study made it clear that Bunker Hill would be hard-pressed to meet the new standard. While fugitive emissions unrelated to smelting undeniably constituted a significant portion of the air lead burden in the Kellogg area,[110] the company's smelting activities were responsible for the lion's share of the total. At all five of PES's air quality monitors in the Kellogg area, lead levels substantially exceeded the federal standard. For Bunker to meet the new requirement, PES determined that "emissions from most smelter sources would have to be reduced by approximately 90 percent."[111] In order to meet the standard at Silver King, the environmental firm estimated that Bunker would have to reduce emissions from most smelter sources "by fractions up to 98 percent."[112]

Bunker Hill responded to the EPA's air lead standard with a familiar stratagem. The company sued, arguing that the rule was unfeasible, capricious, unnecessary for the protection of human health, and highly dangerous to the economic health of the US lead smelting industry.[113] Bunker had numerous co-litigants in this instance, as much of the nation's lead smelting industry joined in support of the LIA suit against the EPA. In a rare courtroom defeat for the company, however, in June 1980 the US Court of Appeals for the District of Columbia ruled in favor of the EPA's 1.5 $\mu g/m^3$ ambient lead-in-air standard. The ruling forced the State of Idaho to develop a lead-air implementation plan for the Kellogg area by 1982. The development of this plan, as with previous implementation plans for sulfur dioxide and particulates, required Bunker to provide government officials with a fair amount of cooperation and detailed technical work. This time, though, the company evidently was not in a very cooperative mood. (Undoubtedly adding to tensions was the fact that by 1980, Gulf Resources had come to view Bunker Hill as more of a liability than an asset and was actively seeking to sell its subsidiary.)[114] In March 1980, Bunker informed the EPA that officials seeking to collect data for implementation of the new lead standard would not be permitted access to company property, ostensibly because the company believed that the EPA should complete its national studies of the lead smelting industry before commencing its research at Bunker Hill.[115] The EPA elected not to force the issue and deferred its study of the Kellogg site. In September 1981, the month after Gulf announced its decision to shutter Bunker if it could not find a buyer for its Idaho subsidiary by year's end, the EPA, acceding to a request by Idaho governor John Evans, agreed to postpone implementation of the new air lead standards to November 1984.[116] Aimed at making Bunker Hill more attractive to prospective buyers by offering "breathing room," the regulatory stay did not, however, lead to a sale of the company.[117]

With the complete shutdown of Bunker's lead smelter and zinc refinery in February 1982, Silver Valley air quality immediately began to improve. Though fugitive dusts from numerous sources continued to offer their bounty of toxic heavy metals to local winds—particularly during the hot, dry months of late summer and early fall—this represented but a fraction of the pollution witnessed when the smelter and refinery were operating. Silver Valleyites literally could breathe easier in February 1982 than they had in decades, but for the more than two thousand Bunker Hill employees who suddenly found themselves unemployed, this likely came as small comfort. Jobless former Bunker Hill workers moved suddenly from the advanced modern society dilemma of worry about the unwanted by-products of industrialization (in this case the health risks of industrial pollution) to the premodern problem of needing to secure adequate food and shelter.

The Yoss Case

In June 1977, the nine children of the families of Bill and Marlene Yoss and Edward and Janice Dennis sued Bunker Hill for $20 million. The lawsuit alleged that the company, by willfully running its smelter at full bore in the wake of the baghouse fire, knowingly caused physical and mental harm to the children, all of whom resided within a mile of the smelter in 1973–1974. The suit maintained that the children, aged between three and seventeen at the time of the suit, suffered from a variety of ailments—from listlessness and stomachaches to impaired cognitive functioning and motor skills—that had been caused by exposure to lead the Bunker Hill smelter had rained down upon them between September 1973 and March 1974. The three Yoss children, Edna, Raymond, and Arlene, all had registered blood lead levels of over 100 μg/dl when they were tested in September 1974. Arlene, of course, scored the highest blood lead level ever recorded, 164 μg/dl.[118] All but one of the six Dennis children had blood leads of at least 40 μg/dl. (the one who didn't was in utero in September 1974), and one tested at between 80 and 90 μg/dl.[119] Many of the children received chelation therapy. Their blood lead levels all declined over time, but as late as 1976 three still measured at 50 μg/dl or higher, with one registering a disturbing 83 μg/dl.[120] Both families lived in neighborhoods that were hard by the smelter during the period of most extreme lead emissions, the Yosses in Deadwood Gulch and the Dennises in Smelter Heights. The fathers in both families worked for Bunker Hill at the time of the baghouse fire, Bill Yoss as a longtime miner.

Bunker Hill responded aggressively. In the wake of the baghouse fire, the company already had faced a similar legal action, a $1 million suit on behalf of the Thomas children. That suit was dropped for reasons that are unclear.[121] Evidently believing that the best defense was a good offense, in the Yoss-Dennis case the company went on the attack, contending that if the plaintiffs' children had medical problems, they were due to unclean homes and neglectful parenting. In court documents, the company's legal team alleged that "Third Party Defendants knew or should have known of the need for cleanliness and the importance of avoiding the contamination of food and the living quarters by said dust, paint or dirt. Notwithstanding said knowledge . . . parents of said minor children, knowingly, negligently and wilfully failed to exercise their duty of care, custody and control . . . and allowed said minor Plaintiffs to be exposed to said soil paint and dust."[122]

In an attempt to lay the blame for any medical problems that the Yoss or Dennis children—and by implication any and all Shoshone County children—might have suffered due to the area's heavy lead burden, Bunker held that it was the responsibility of parents, and particularly of mothers, to keep their houses clean and their toddler-aged children from eating yard dirt. While the company did not deny that dust in the Kellogg area was high in lead content or that ingesting that dust was potentially bad for one's health, it laid full responsibility for the prevention of lead-related children's health problems squarely on the shoulders of parents. Good parents, implied the company, understood how to maintain clean homes and keep their children healthy. Bad parents either lacked the knowledge or ability to properly safeguard their children in the Kellogg area's leaded environment.

During the more than four years that the Yoss case lasted, the nine plaintiff children submitted to numerous batteries of physical and psychological tests at medical facilities from Chicago to Seattle. While none of the children showed signs of severe mental retardation or were fully disabled, all evinced significant signs of lead exposure. A number of them complained of severe headaches and stomachaches. Clear evidence of exposure to large quantities of lead was found in some of the children's bones, where lead lines were visible.[123] Many of the children were diagnosed with hyperactivity and an inability to focus on tasks at hand. Although one was identified as mildly retarded, the rest were found to be in the average-to-low-average range for intelligence.[124] Two of the children, however, experienced unusually large drops in IQs, of twenty-seven and twenty-one points respectively, between their November 1976 and March 1979 testings.[125]

One of the younger children was found to have severe difficulties with "processing problems" that led the child to experience a good deal of "internal confusion." This, in turn, placed the child under a burden of emotional distress. The diagnosing psychologist advised offering this child "special learning resources" and "family therapy" to help cope with these ongoing difficulties.[126] Another child, after passing the gross motor section of the Cott test of motor impairment, failed the test's fine motor section, a highly unusual occurrence. The overseeing doctor concluded that this result likely was due to the poor condition of the subject's median and radial nerves.[127] Impaired nerve functioning is a principal effect of lead intoxication. A majority of the children, and all who were under the age of four at the time of the baghouse fire, were found to have moderate to severe visual-spatial impairments. Dr. Agnes Lattimer, who examined four of the children at Cook County Hospital in Chicago in late 1976, diagnosed all as "showing some signs of central nervous system damage," and as currently exhibiting "clinical manifestations of exposure to lead in toxic amounts."[128] Lattimer recommended additional chelation therapy for all four children.[129]

Most of the children were having serious difficulties in school, struggles their doctors predicted would continue due to the organic impairment the children had suffered from their experience with severe lead intoxication. Addressing the likely need for long-term special educational help for some of the children, doctors referenced the "organic substrate" of lead in the children's bodies that likely would continue to interfere with their development for years to come.[130] Whether lodged in soil or in the bones of children, lead was a heavy, obdurate substance that was not easy to remove. The doctors also diagnosed a wide range of learning problems that included brain impairment, poor visual-motor integrations and "bizarre reversals," adjustment problems, extremely low math and language comprehension scores, and an absence of the ability to recognize and deal with problems.[131]

While all of the Yoss and Dennis children showed effects from their lead exposure, some were clearly more damaged than others. The older children generally appear to have been less severely harmed than were the younger ones. According to the doctors, between two and four of the older children were not significantly impaired by their brush with lead.[132] Still, a lifetime earnings model developed for the trial depicted reductions in the children's incomes, based upon their injuries, of between 50 and 90 percent. This amounted to projected lifetime losses of between $117,777 and $454,416.[133]

Although it is extremely difficult, if not nearly impossible, to prove with certainty that an individual illness was caused by any one factor—witness the nearly unbeaten track record of the tobacco industry in defending against medical claims—the extensive medical records of the Yoss children make a strong case. The records indicate that ingested lead did real, long-lasting, physical and mental damage to the children. In the case of some of the younger children that damage appears to have been particularly severe.

After years of depositions, evidentiary disclosure, expert witness testimony, medical probing and legal wrangling, the case of *Edna Grace Yoss et al. v. The Bunker Hill Company et al.* came to trial before US District Judge Roy McNichols in Boise, Idaho, in September 1981. The trial lasted six weeks before Bunker Hill and the plaintiffs reached an out-of-court settlement. The complex deal, formally ratified by Judge McNichols at the end of January 1982, guaranteed the seven younger Yoss and Dennis children between $7.1 million and $8.8 million over the course of their lifetimes ($18.5 million and $22.8 million in 2011 dollars), depending upon how long they lived.[134] Evidently judging the harms done them to be relatively minimal, the settlement awarded the two oldest children only $10,000 each. Monthly annuity payments began immediately and the children were to collect large lump-sum payments in between ten and thirty years.[135] While the settlement undoubtedly granted the seven younger Yoss and Dennis children a significant amount of monetary wealth, it is uncertain that this (or any) material compensation could serve as an adequate substitute for the loss of health and personal well-being that the children appear to have suffered. The Yoss and Dennis children are classic examples of modernity's uneven "distribution and growth of risks" and the consequent rise of "social risk positions."[136] Growing up in poor neighborhoods hard by the Bunker Hill smelter placed children like the Yosses and Dennises in social risk positions unlike those of children growing up in nearby (but significantly less unhealthy) Cataldo or Wallace. While these seven children received significant compensation for the health damages they sustained, the same was not true for their Deadwood Gulch, Smelter Heights, and Smelterville peers who were placed in similar risk positions.

Proof of Harm

While the Yoss case settlement provided some public evidence of harm done to Silver Valley children by the lead outbreak of 1973–1974, it took another fifteen years for more definitive proof of damage to emerge. In 1996, the federal Agency for Toxic Substances and Disease Registry (ATSDR) published a report on the long-term health of Silver Valley children who were exposed to high doses of lead in the wake of the baghouse fire. The study tracked 1,466 adults who lived in the communities of Kellogg, Smelterville, Wardner, Page, and Pinehurst in 1974–1975 and who were between nine months and nine years of age at that time.[137] Researchers interviewed 917 and tested the nervous systems and the levels of bone lead of 281 of the former Silver Valley children.[138] As a control group, the study used 754 Spokane residents between nineteen and thirty years of age, 287 of whom participated in the tests. The test results revealed striking differences between the Silver Valley group and the control group. The former group suffered from significantly higher rates of a long list of disorders that included infertility, learning disabilities, sleep disorders, high blood pressure, anemia, arthritis, depression, anxiety disorders, memory loss, and trouble concentrating. The Silver Valley cohort also performed worse on nerve conduction tests and had reduced hand-eye coordination and grip strength, diminished nerve sensitivity, and limited vocabulary.[139] The retinue of health problems found in the Silver Valley group more than twenty years after the baghouse fire, coupled with the leaded group's heightened rate of disorders as compared with the control group, offered strong evidence that, sadly, significant numbers of children who lived within a few miles of the Bunker Hill smelter in

1974–1975 had suffered long-term health damage from their extreme lead exposures of that era.

Some of the children who grew up amid North America's worst environmental lead outbreak, now middle-aged adults, believe they were irreparably harmed by the experience. Calling themselves "the leaded," a group of them gathered in 2010 in Smelterville to tell their story and urge Silver Valley parents to have their children's blood lead levels tested. Members of the group complain that they have suffered physical and emotional problems since the mid-1970s. Jeannie Stancik, a forty-eight-year-old Kellogg resident, said, "I tell everyone I'm leaded. That's how I deal with it." According to Cass Davis, "People don't want to know the truth. I flunked the first grade and was sent to be tested for retardation."[140] Sadly, there are many more such cases. Lori Wilson grew up 250 yards from the smelter and had a blood lead level of 98 μg/dl in 1974. As a twenty-five-year-old in 1996 she could not read, had been unable to complete high school, and was living with her parents. Dain Milholand, a Smelterville baby in 1974, failed his driving education exam eight times before finally passing.[141] As Bunker Hill partisans have pointed out, however, not all children who underwent extreme lead exposures suffered later difficulties. Kyle Wombolt, for example, a five-year-old in 1974 whose childhood home was next to Lori Wilson's, played high school football, graduated at the top of his class from the University of Idaho law school, and became a Wall Street lawyer. A number of other similarly exposed children also went on to college and have had successful careers. Because of differences in diet, exercise, metabolism, genetics, and other factors, high lead exposures did not affect all Silver Valley children in the same way.[142] A common refrain in Kellogg has been that such successful, healthy children came from households that practiced good hygiene, where children were required to bathe and wash their hands frequently, whereas children who were sickened by lead must have come from poor homes where cleanliness was not the rule. In reality, however, the area's lead health situation has never been as clear-cut as proponents of hygiene as a cure-all contended. As Jerry Cobb, the leading Shoshone County official on lead health issues since the mid-1970s, says, "I knew of families with floors one could eat off who had kids with high blood leads. The smelter didn't discriminate."[143]

Coming Up for Air

In the wake of the lead health epidemic and its attendant bad publicity, Gulf-Bunker officials realized the need to take some remedial environmental measures. In addition to footing most of the bill for the Shoshone Project and hiring a public relations firm, in October 1974 the company hastened its efforts to purchase homes and lots in the neighborhoods of Deadwood Gulch, Smelter Heights, and

Silver King—all located within half a mile of the smelter. After purchase, Bunker razed the homes to prevent anyone from living in this most heavily leaded portion of the Valley. By September 1975, the company had purchased sixty-two of the area's seventy-seven homes.[144] In October 1974, for the first time in about twenty years, Bunker shut down the smelter for three days of general cleaning and repair.[145] The company also began providing clean topsoil to locals who wanted to try their hand at growing a lawn, and purchased street sweeping equipment to limit the dust on the streets and playgrounds of communities near the smelter.

For a number of years Bunker also funded the annual, voluntary blood testing of children living within 2.5 miles of the smelter. Overall mean blood lead levels decreased dramatically in 1975 and continued to decline for the next few years. Mean blood lead for children living within a mile of the smelter declined consistently each year, going from 71.6 μg/dl in 1974 to 40.0 μg/dl in 1977. In 1978, however, they increased slightly, to 42.0 μg/dl.[146] For children living between 1 and 2.5 miles from the smelter, blood lead levels declined only very slightly from 1974 to 1975, moving from 58.9 to 58.4 μg/dl, but saw consistent decreases the following two years, reaching the "safe" level of 35.5 μg/dl in 1977.[147] At the end of the 1970s, however, mean blood and air lead levels rose again.[148] Due to the rapidly declining number of subjects, however, the blood lead results were not as significant as they might have been. Participation rates began to decrease with the blood draw of 1976, when only 193 children living within 2.5 miles of the smelter showed up versus the 623 children who were tested the previous year. By 1979, a mere 28 children participated.[149]

In 1980, responding to pressure from CDC director Dr. William Foege, and Bunker's longtime thorn in the side, Philip Landrigan, the State of Idaho and Bunker Hill renewed their efforts to study and address lead health issues in the Kellogg area. In a letter to the Idaho Department of Health and Welfare, Foege encouraged the agency to accept the CDC's offer to conduct a comprehensive, follow-up investigation of the health of the "leaded" children studied in 1974–1975. Aware that the agency was contemplating accepting financial support from Bunker Hill to do its own study of the issue, Foege urged the state officials to reconsider. The CDC director wrote Dr. Edward S. Gallagher, Idaho's State Health Officer, to stress both that a company-financed study would lack the full scientific credibility of a CDC study and that the severity of Kellogg's lead health problems warranted the level of seriousness and professional standards that only the federal agency would bring to bear. Foege wrote that "although we have no wish to be alarmists, we must point out to you that . . . the Silver Valley is the site of the worst community lead exposure problem in the United States."[150]

At about the same time, national political activity and news coverage followed the EPA's decision to implement its new ambient air lead standard. In December 1979, US Representative Henry Waxman (D-CA) held hearings on the air lead standard before the House Subcommittee on Health and Environment. Among those Waxman called before his subcommittee were Janice Dennis and Philip Landrigan. In her testimony, Dennis, mother of six of the children involved in the *Yoss v. Bunker Hill* case, painted a grim picture of living conditions near the smelter and of an uncaring company that presided over that environment. Dennis testified, "There was no grass, and flowers didn't grow; there weren't even weeds." Dennis also noted the difficulty of keeping domestic animals alive in Smelter Heights—the family's five puppies, and assorted chickens, rabbits, and turkeys all had perished.[151] In his testimony, Landrigan suggested that the State of Idaho was dragging its feet in conducting, or allowing the Public Health Service to conduct, needed follow-up studies of children affected by the 1973–1974 lead outbreak.[152] Landrigan recently had stirred the pot by disputing the Shoshone Project's official conclusion that children's health had not been damaged and by distancing himself and the CDC from the project. The doctor had said that the Shoshone Project glossed over more troubling aspects of its own findings and that CDC officials such as himself had "had as little to do as we could with the Shoshone Lead Project."[153] In addition to their call for the State of Idaho to permit the CDC to conduct a comprehensive lead health review in Shoshone County, both Foege and Landrigan asked that the state heed the Shoshone Project's call to perform neurological testing of the most heavily leaded of the previously tested children.

In spite of the high-profile pressure, Idaho's Department of Health and Welfare declined the CDC's offer to manage and coordinate a renewed lead health study. Once again, despite the CDC's warnings that doing so would compromise the study's credibility, Idaho permitted Bunker Hill to largely bankroll the project, this time to the tune of $125,000.[154] The lead health study conducted in 1980–1981 included three components: blood lead monitoring of children between the ages of one and twelve living in the areas of Smelterville, Kellogg, and Wardner, efforts by a community health team to conduct lead health educational activities and to identify and reduce lead exposure in high-risk family and community settings, and neurological testing of Shoshone County children whose blood leads had measured higher than 80 μg/dl and of children born between October 1972 and March 1975 who lived within a mile of the smelter during their first year of life.[155]

The new, better-publicized round of blood testing, which paid children $15 apiece, attracted far more participants than had recent efforts. The tests also displayed more favorable blood lead trends. Where only 27 children had been tested

TABLE 5.3. Mean blood lead levels (μg/dl) for children living 1–2.5 miles from the smelter

1974	1975	1978	1980	1981	1982	1983
49	38.7	35.5	31	27.2	21.2	< 18

Note: The 1980 blood testing was done in October. In other years the testing was performed in August.

Source: Idaho Department of Health and Welfare, *Kellogg Revisited—1983 Childhood Blood Lead and Environmental Status Report*, p. 17–18, MG 413, Box 253, Folder 4716, Bunker Hill Records.

in 1979, the October 1980 screening attracted 147.[156] Mean blood lead levels within a mile of the smelter that had measured 42 μg/dl in 1978, by October 1980 had dropped by over 25 percent to 30.5 μg/dl. Blood lead levels also declined significantly for the area between one and 2.5 miles from the smelter. Both trends continued over the next few rounds of testing, with relatively robust numbers appearing for blood draws and decreasing quantities of lead in children's bloodstreams. The shutdown of the smelter in early 1982, eliminating the valley's primary source of airborne lead, accentuated the positive trend, with children's blood lead levels further declining in testing done in August 1983 (see Table 5.3).

Not surprisingly, following the closure of the smelter and zinc refinery, ambient levels of air lead declined dramatically. Smelterville and Kellogg air came into comfortable compliance with the 1.5 μg/m^3 standard, likely the first time that Silver Valley air had been so free of lead since 1960 when a strike idled the facility for seven months. Since lead is less easily leached from the soil than from the air, however, soil concentrations of the metal remained high. Soil lead means around area homes measured as high as 6,372 ppm, and individual levels as high as 18,400 ppm were recorded in 1983.[157] Soil lead levels, in fact, were comparable to those found in 1974, indicating that cleansing area soils would be an even more daunting challenge than cleaning the local air and water.[158] Lead concentrations in household dust, while still elevated, decreased to a fraction of their 1974 concentrations (see table 5.4). The primary threat to Silver Valley children's health, the Bunker Hill smelter, was gone. Secondary threats, however—most particularly high quantities of lead in soil and household dust—remained.

The 1980s witnessed significant progress in the lead health of Silver Valley children. By 1989 about half of the children tested measured under 10 μg/dl of lead in their blood.[159] During the 1990s hundreds of children residing in "the Box," a seven-by-three-mile area centered on the old Bunker Hill smelter that in 1983 became an EPA Superfund site, went under the needle each year.[160] In 2000, the Panhan-

TABLE 5.4. Mean household lead dust levels (ppm) at Smelterville

1974	1983
10,000	3,994

Source: Wegner, *Shoshone Lead Health Project*, 53; and Idaho Department of Health and Welfare, *Kellogg Revisited*, 25.

dle Health District reached its goal of bringing the blood lead levels of 95 percent of these children below 10 μg/dl—the blood lead level threshold the CDC has considered "unsafe" since 1991.[161] These blood lead reductions undoubtedly were abetted by the EPA's substantial efforts to reduce soil lead levels in the valley. Commencing in 1989, the agency oversaw the digging up of household yards, city parks, and roadside shoulders. By July 2008, when the EPA officially announced completion of the project, approximately three thousand residential and other properties had seen their soil dug out up to a depth of four feet and replaced with clean topsoil.[162] (The contaminated soil was placed in the EPA's massive, "sealed" earthen sarcophagus that sits between Kellogg and Smelterville.) With this, the problem of lead in the soil of communities nearby the old Bunker Hill smelter had been largely addressed. Significant strides also have been made with regard to house dust. The EPA judged the amount of lead in the valley's house dust to be below the standard of concern in 2010.[163] Reflecting the diminishment of bioavailable lead in the local environment, by 2007 Silver Valley children's blood lead means were in line with the national average of 2 μg/dl.[164]

Due to the amount of lead and other toxic heavy metals that were spewed into the atmosphere of the Silver Valley over the sixty-five-year life of the Bunker Hill smelter, there will long remain ample reason to keep close watch on the health of area people, especially children. Although the lead health improvement of Silver Valley children in recent decades is encouraging, it should be remembered that there remain a significant number of adults, raised in the area, who suffer from a wide variety of serious health problems owing to their lead exposures as children in the 1970s.

Western historian William G. Robbins is among those who have noted that the tradeoffs of industrial modernity have been spread unevenly over the United States. Simply stated, some places have suffered far greater environmental damage than others. Some places have remained far poorer than others, and the residents of some communities have been sickened far more than others. Discussing the American West, Robbins suggests that the region's "hinterlands" have tended to suffer from economic neglect and environmental despoliation while its urban centers (e.g., Denver, Phoenix) have thrived. He points to Kellogg and the Coeur d'Alenes as a prime example. Robbins writes, "Begin with Kellogg, Idaho, where more silver has been mined in the last century than any other place in the world. The famed Silver Valley, with its ribbon of mining towns . . . was once alive with the vital stuff of working-class life. Today, the barren hillsides around Kellogg and the huge slag heap at one end of town are the singular reminders of those industrial successes."[165] While Kellogg and the rest of the Coeur d'Alenes undoubtedly garnered sizable proceeds for their industrial efforts—money that once went toward building strong, vibrant communities—the region today would hardly be confused with a wealthy, "pristine enclave" such as an Aspen or a Sun Valley.

If asked to name the places in the United States that have suffered the greatest damage to environmental and human health, many Americans probably would suggest Love Canal, New York, a community that became infamous nationally in the late 1970s for the toxic waste dumping that precipitated a major environmental health crisis. Others likely would mention former nuclear weapons arsenals such as Rocky Flats, Colorado, and Hanford, Washington, well known for their long and expensive cleanups. Those more attuned to contemporary environmental issues and discussions might mention Louisiana's "Cancer Alley," Toms River, New Jersey, or Flint, Michigan.[166] Relatively obscure throughout its history, despite its importance as a mining and smelting center and its status as home to the greatest environmental lead health crisis in the nation's history, it is unlikely that Kellogg, Idaho, would receive many mentions in such a poll.

CHAPTER 6

ON LEAD'S FRONT LINES

Bunker Hill's lead and zinc plant workers—the company employees who experienced the highest exposures to toxic heavy metals—lie at the heart of this chapter. These workers literally put their bodies and health on the line for the job. Many suffered health consequences as a result of their years of work in the plants.[1] Subjected to high concentrations of airborne lead and zinc at work, many plant workers returned to their homes only to be again surrounded by air, dust, and dirt containing among the highest levels of those metals to be found anywhere in the United States. Bunker's lead and zinc plant workers, however, were far from passive proletarian victims of environmental insult. This chapter illuminates their struggles—voiced principally through Local 18 of the Mine-Mill union and later through Local 7854 of the United Steelworkers—to improve working and health conditions at the plants. As times changed, so did government, worker, and corporate attitudes toward acceptable levels of workplace pollution, although often at differing rates. The shifts in federal policy that accompanied the environmental turn of the late 1960s and early 1970s, and the struggles those policy changes precipitated on the shop floor of Bunker Hill's plants, are laid out in the following pages. So too are the efforts of Bunker workers to suspend hard-won federal health rules in their near-desperate attempt to attract a buyer for their failing company in the early 1980s.

It almost goes without saying that neither Bunker Hill managers nor, certainly, smelter employees themselves *wanted* workers to be sickened or injured on the job. But, as environmental historian and legal scholar Arthur F. McEvoy notes, such injuries to workers, like harm to nature, were long viewed as simply unavoidable, albeit unevenly shared, costs of development. McEvoy writes, "Like pollution, occupational injury and disease are unwanted consequences of industrial development. They are part of development's social cost, typically falling at random to a diffuse body of victims and figuring only marginally, if at all, in the management decisions that prepare the ground for them."[2] This chapter seeks to place the health of Bunker Hill Company smelter and zinc refinery workers, or what has been called "the nature within us," at the heart of a story that depicts the shifting workplace ecologies of the Bunker Hill lead smelter and zinc plant. Bunker Hill workers knew these complex ecologies intimately through their labor within them.

McEvoy has called for technological-ecological studies of work environments "of which the worker's body is the biological core," and he has characterized these highly engineered, profit-oriented spaces as "the product of ongoing, reciprocal interaction between biology, political economy and consciousness."[3] This chapter will explore some of the ramifications of these complex, reciprocal interactions upon the constantly shifting workplace ecologies of Bunker Hill's lead smelter and zinc plant.[4]

Risk is a necessary aspect of industrial modernity, but it is rarely distributed evenly through society. Within industry the physically riskiest jobs of course will be found on the shop floor, not in the executive suite. This was certainly the case for the Bunker Hill Company, where miners, along with lead and zinc plant workers, assumed the greatest share of the risk within the company of serious injury from accident or health damage from exposure to aerosolized heavy metals. Although Bunker workers undoubtedly were paid somewhat higher wages and benefits due to the dangerous and unhealthy nature of their jobs, these payments likely fell short of covering the costs in lost wages and damages that workers sustained for their injuries and health problems.

Leaded Workers: An Old Problem

From the beginning of smelter operations in July 1917, top Bunker Hill officials knew that the problem of "leading" workers at the new facility was one that had to be addressed. By this time the work of pioneering American industrial hygienist Alice Hamilton was well known. A more immediate concern for Bunker Hill managers, however, was simply securing an adequate workforce at the smelter. The smelter's at best unpleasant working environment, however, handicapped the company's recruitment efforts. As Bunker Hill manager Stanly Easton complained in a 1917 letter to company president F. W. Bradley, "The labor situation remains very unsatisfactory, more so at the smelter than the mine, few men are applying for work . . . the intense heat and the discomforts of smoke and dust around the new equipment at the smelter has caused the rapid falling away of former employees."[5]

In June 1918, Dr. Royd R. Sayers, a surgeon with the US Public Health Service, spent a few days visiting the smelter, investigating worker complaints of occupational disease on behalf of the Bureau of Mines.[6] In his report, Sayers judged that silicosis was a greater threat to smelter workers than lead poisoning, but still pointed to a number of areas in the plant that threatened workers with lead poisoning.[7] Among these was the tapping floor, where Sayers concluded, "The fumes . . . are very bad and extremely dangerous from the standpoint of lead poisoning."[8] The public health official also determined that the existing method of removing dust from the Cottrell precipitator "unnecessarily exposes the workmen," and that the "droppings from the Dwight and Lloyd machines . . . was creating quite a little dust as it fell onto the conveyor."[9] The dust, Sayers also believed, posed a threat to smelter workers.

Succinctly noting that "whoever works in or about a lead smelter may become leaded," Sayers stressed the relative importance of cleaning the plant environment over encouraging improved worker hygiene, while encouraging the pursuit of both.[10] The industrial health expert's recommendations included settling leaded dust with water wherever possible, providing physical examinations to smelter employees every six months, with "especial attention being given to signs or symptoms of lead poisoning," and notifying workers of the dangers of lead poisoning along with the best means of avoiding it.[11] Among Sayers's nuggets of hygienic wisdom for workers were: "Avail yourself frequently of the shower baths furnished by the Company. . . . When passing to the leeward side of fumes, hold your breath as much as possible. . . . Where fume or dust is abundant, tie a handkerchief around your nose, or better wear one of the respirators provided by the Company. . . . Put a little plain vaseline in your nose at the beginning of your shift," and finally, "Do not allow yourself to become constipated."[12] Despite the health official's conclusion that control of dust and fumes at the plant was more crucial than improving worker hygiene, however, Bunker opted to emphasize the latter.[13]

The company stood both to save money and also to lower its potential legal liability by laying the principal burden of responsibility for worker health safety at the lead smelter on the workers themselves. In so doing, the company remained in keeping both with the industry standard of the day and with the prevailing legal doctrines of "assumption of risk" and the "fellow servant rule."[14] Still, as Easton confided in a letter to Bradley in 1924, lead poisoning cases at the smelter, "while not as frequent or as severe as formerly still occur."[15] Although to Bunker's relief, Idaho (also in 1924) ruled lead poisoning to be an occupational disease not covered under the state's workmen's compensation law, the company decided to offer compensation to "worthy employees" whom the company judged to have lead poisoning.[16] Nonetheless, Easton and Bradley still worried over the possibility of a successful lawsuit by "leaded" employees who were unsatisfied with the amount of offered compensation. To defend against this threat, Easton advanced the posting of "repeated warnings" and "conspicuous notices" at the smelter that warned of the dangers of working in a leaded environment and gave instructions on "effective ways of preventing lead poisoning."[17] Thus insulated by its warnings and advice, Bunker Hill bargained on employing the nineteenth-century legal doctrine of assumed risk to avoid liability for any lead-poisoning maladies that smelter workers might contract.

Though unwilling to expend significant resources to remove leaded dust and fumes from the working environment, Bunker did try various "technological fixes" in hopes of eradicating the problem of leaded workers. The company first employed the Clague electrolytic system in 1919, urging all workers to take the treatment regularly.[18] The Clague treatment involved the immersion of hands and feet in a warm saline solution, through which a weak electrical current then was passed.[19] Hour-long treatments for up to forty men at a time were conducted at Wardner Hospital.[20] The system did not witness much success, however. Bunker hoped for better luck with its highly touted solarium, opened in September 1929 at a cost of ten thousand dollars.[21] Company officials believed the artificial sunlight of the solarium would counteract the effects of lead on workers' bodies, and that it offered additional health benefits to workers and their families, particularly during the long, overcast, north Idaho winters.[22] Capable of treating 300 people an hour, during 1930 the solarium provided 3,845 free treatments to Bunker Hill workers, their wives, and children.[23] There are, however, no reports of "cures" from the solarium treatment.

Health Care for Miners and Smeltermen

Given the dangers of working in the leaded smelter environment or in the mines of the Coeur d'Alenes, it is not surprising that access to adequate health care has long been a priority of the area's miners and smelter workers. In the early days of the district, mining companies typically deducted one dollar from miners' paychecks to support medical care through a mutual aid fund.[24] Bunker Hill maintained an in-house physician in the 1890s–1900s, Dr. Hugh France, who treated many company workers. However, miners often resented their lack of control over their medical care. In 1892, the Miners Union demanded control over the mutual aid fund. The union wanted to use the money to build and run a hospital that would specialize in mining injuries.[25] Rather than let inquisitive miners delve into the operations of the mutual aid fund, Bradley and Easton dissolved it in 1909 and turned their support directly to the new Wardner Hospital, built by the company on Bunker land in 1907 at a cost of approximately $12,000. Initially administered by Dr. France, it functioned essentially as a company hospital until the completion of the public Shoshone County Hospital in Kellogg in 1958.[26] During the 1930s Bunker continued to offer health care to employees through Wardner Hospital. By the end of the decade, the hospital, with three doctors and a dozen nurses, offered forty beds and boasted "modern X-ray and surgical apparatus."[27] Unlike most hospitals, however, Wardner provided a specially trained industrial health nurse and a public health nurse, both of whom made "visits to employees homes when there might be an unreported illness or other need." The hospital also provided an outpatient department offering treatment for injured workers at all hours. Annual physical examinations for Bunker employees were conducted at the hospital, during which "a close check [was] maintained on the status of silicosis and tuberculosis throughout the plant." The company also used the examinations to determine when older or sicker workers needed to be shifted to less hazardous jobs or workspaces.[28]

Changing Times

In 1917, the Idaho legislature enacted a Workmen's Compensation Act that was modeled closely on that of its neighboring state, Montana. While the law was somewhat more robust than Bunker Hill executives would have liked, they nonetheless were relieved that the Idaho act was not as strong as Washington's and Oregon's.[29] Following the provisions of the new law, Bunker entered into a contract with Wardner Hospital to offer medical and surgical services to injured company employees. To compensate the hospital the company paid twenty-five cents per month for each employee and employees each paid one dollar per month, the maximum they were permitted to contribute under the law.[30]

In the 1930s, shifting political winds gave control of the Idaho state legislature and the governorship to New Deal–leaning Democrats. Under new pressure from Boise and Washington, DC, Bunker Hill agreed to implement changes at its lead smelter that had been advocated by Dr. Sayers nearly twenty years earlier.[31] The company sent one of its engineers, Harold Carlson, to the Air Hygiene Foundation of America, in Pittsburgh, for specialized training in dust-control methods. In the latter part of 1936 Bunker adopted a program that consisted of physical examinations for all employees at the time of hiring and at regular intervals thereafter; routine sampling of air in the smelter, zinc refinery, and in the mines; installation of equipment for reducing dust in these facilities; and periodic sampling to test the effectiveness of dust-controlling systems. Bunker employed an Impinger dust-collecting apparatus and an Owens jet dust counter to collect dust from the smelter and other facilities, and a Zeiss circular konimeter with a built-in microscope to test the dust's contents. To decrease the amount of metals-laden dust in the facilities, the company installed water sprays and exhaust systems. In an effort to protect at least some workers, "dust-producing machinery was isolated to pre-

FIGURE 6.1. Bunker Hill workers line up on payday, late 1950s.

vent escape of dust throughout the entire building." This measure, however, likely concentrated dust in certain parts of the plants, where workers would have been subjected to greater concentrations of lead and other heavy metals. Under the new policies, however, workers were to be provided with respirators "when exposed to excessive amounts of dust."[32]

The United States' entry into World War II brought major changes to the Bunker Hill Company and the Coeur d'Alenes. Heightened demand for area metals such as lead and zinc spurred renewed activity at the area's mines and in Bunker's smelter and zinc refinery.[33] In addition, the federal government's role in the mining district became decidedly more pronounced. Under the auspices of the National War Labor Board's Non-Ferrous Metals Commission, 50 percent of the district's zinc and 15 percent of its lead production were set aside for the war effort.[34] Even more significant, however, was Bunker Hill's reluctant decision to recognize union representation of its workers for the first time in company history.[35] Following the National Labor Relations Board's (NLRB) ruling that Bunker's zinc plant was an appropriate bargaining unit, workers at the plant voted in October 1941 to organize under the International Union of Mine Mill and Smelter Workers (IUMMSW).[36] In late December, following Pearl Harbor and US entry into the war, the company signed a contract with the newly unionized plant. The following year, the NLRB ruled that the remaining Bunker Hill workers had a right to organize. In May 1942, workers at the company's mine and lead smelter also voted to join the union, and so became (along with their zinc plant brethren) Local 18 of the IUMMSW.[37] Although Bunker management hoped to roll back union recognition and wartime labor concessions following the end of the global conflict, that did not occur. The company would remain a closed shop for the next forty years of its existence.

During World War II Bunker Hill faced a chronic shortage of workers. Taking advantage of the heightened leverage, Local 18 asserted its position on a number of issues, including workplace conditions. Meeting high production goals with relatively few workers meant a production "speedup" for employees. This, in turn, brought worsened dust and gas conditions at the smelter and refinery.[38] Workers at the smelter baghouse argued for higher wages on the grounds that they should be provided added compensation for enduring such hardships. Likewise, granulating men called for a raise due to the great amounts of dust created by the crusher, ore-unloading crew members demanded an extra seventy-five cents per shift due to dust and gas conditions that they maintained were the worst of any at the plant, and the union advocated time-and-a-half pay for those doing the extremely dirty job of flue cleaning.[39] During the war period the union viewed particularly unhealthy working conditions as grounds to demand higher pay for affected lead and zinc plant workers, but stopped short of insisting that the company remedy the conditions. The plants' internal environments, Bunker labor and management seem to have tacitly agreed, could remain relatively dirty and unhealthy. Instead of battling over issues of pollution within the plants, Local 18 and the company wrestled over pay levels for particularly unhealthy jobs and the provision of enough respirators to workers performing those jobs.

The Postwar Era

Bunker Hill experienced a major change of leadership in the mid-1950s with the retirement of longtime company president Stanly Easton and his replacement by John D. "Jack" Bradley, the son of former company president F. W. Bradley.[40] The forty-five-year-old Bradley's accession represented a generational shift in leadership. The younger man brought a number of new priorities to the job—among them modernizing operations and increasing worker productivity at the seventy-year-old company. The latter concern would bring the company into direct conflict with Local 18. What Bunker management viewed as "increasing worker productivity," workers saw as submission to the dreaded "speedup," with all the attendant health and safety issues that entailed.[41] Viewing time spent in high concentrations of fumes and gases as central to the problem of lead intoxication, the union demanded larger crews and more relief men to enable particularly dirty jobs to be done more quickly.[42] In January 1954, Local 18 posted a warning about lead health dangers at the smelter on the facility's bulletin board advising workers that "there have been two men leaded at the smelter due to the speedup since our conference with the smelter management. . . . We hope that this will alert the membership to the hazards that exist here."[43] Workers also attempted to curb the negative health effects of the speedup through changes in contractual bargaining language. A large-scale smelter grievance in 1954 attempted to limit the size of work crews and the pace of work. The company, however, successfully pushed to increase production first and deal with health and safety concerns later.[44]

In early 1957, Bunker Hill announced a $15 million, five-year plan to expand and modernize its plant and equipment.[45] While many in Kellogg and neighboring communities undoubtedly were encouraged by the economic benefits that presumably would result from Uncle Bunker's growth, some expressed concerns. It was well understood by locals that expansion of the lead smelter and zinc plant meant increased production, which, in turn, meant the venting of more metals-laden smoke and fume into the area's air. The Kellogg City Planning Commission inquired about company efforts to curb the "smoke problem." In response, President Bradley presented the community with a classic economy-versus-environment tradeoff, saying, "I wish to emphasize that the production of gas is a necessary result of the operations of our reduction plants. As we enlarge these operations, more men will be employed resulting in more business activity in Kellogg—but more gas will be produced."[46] According to Bradley, Kelloggers could have prosperity or clean air, but not both. Bradley's arguments carried the day. Jack Bradley's bald "price of progress" argument and the community's muted assent evidences that, in 1957, Kellogg was still living in a stage of industrial development that Ulrich Beck refers to as *classic industrial modernity*, wherein the material benefits of industrialization outweigh its risks and other "side effects" for society, and "the production of risks . . . can be legitimated as 'latent side effects.'"[47]

More so even than their neighbors, smelter and zinc refinery workers faced the burdens of the increased pollution caused by the plant expansion of the late 1950s. Although the company, in efforts to combat "dangerous lead gases," installed twenty-four fans in the smelter roundhouse and twenty-two in the lead refinery, silver refinery, and Dwight & Lloyd roasters, lead refinery workers still complained of "intolerable conditions of smoke." Blast furnace workers complained that their health and safety were "endangered by the vast amount of smoke, gases and noxious fumes in and around their work area."[48] During the tense contract negotiations of 1959, Local 18 argued that Bunker had neglected worker health and safety. More specifically, the union complained that by failing to take advantage of opportunities to clean the smelter's blast furnace during times of routine maintenance, Bunker subjected workers to excessive quantities of toxic fumes. Local 18 used this to illustrate that unless compelled by contractual language to adequately safeguard employee health and safety, the company could not be relied upon to do so.[49]

In the battle for local public opinion during the epic, seven-month Mine-Mill strike of 1960, the union portrayed Bunker Hill as, among other things, a despoiler of the local environment. Local 18's radio broadcasts noted "how much our children's throats clear up each day they do without smelter smoke," and union members claimed that they were able to get suntans now that smelter fumes were not blocking the rays.[50]

The strike ended with a vote that narrowly rejected Mine-Mill in favor of a new union organization, the Northwest Metal Workers Union (NWMW). Despite the fact that the McCarthy era was long over in most of the country, international communism became a central issue in the contentious labor conflict. For years Bunker Hill had made political use of the fact that members of the IUMMSW's national leadership either were or had been members of the Communist Party, and alleged that local Mine-Mill leaders also were Reds.[51] Anticommunism, a strong force in the district since at least August 1951, when an allegedly pro-Soviet talk on local radio station KWAL prompted popular outrage, proved potent in the resolution of the strike of 1960.[52] The charge of communism became a weighty burden for Mine-Mill to carry, and its rival, the NWMW, portrayed itself as strongly anticommunist. Following its ascent, the NWMW refused to allow membership to much of the former leadership of Local 18 on the grounds that those unionists were either communists or communist supporters.[53] This group of twenty-five militants became a vocal opponent of the NWMW, first dubbing itself the "Reject Club," and later adopting a more formal name, the Committee for United Trade Union Activity.[54]

Signs of Discontent

Since 1954 Bunker had conducted regular urinalysis screening of smelter workers to determine whether individuals were leaded. Depending upon their presumed levels of exposure within the smelter, employees were tested monthly, quarterly, or semiannually.[55] Smelter employees whose urine lead levels reached 300 micrograms per liter (μg/L) were referred to the doctor.[56] Workers who tested high for lead were transferred to another, presumably more lead-free job. Also, smelter workers in particularly dusty or dirty parts of the plant were required to wear respirators. Despite these measures, in the militants' view work speedups at the smelter were resulting in greater numbers of lead-poisoned workers.[57] The Reject Club maintained that eighteen of twenty-three workers in one department showed elevated lead levels. Asserting that this was indicative of the lead health situation throughout the smelter, the group called for a relief man on all shifts so workers could take breaks from the dust and fumes.[58]

In June 1961 the Committee for United Trade Union Activity sent an open letter to several state and federal officials, including Idaho governor Robert Smylie. The letter maintained that "the incidence of lead poisoning [at the Bunker Hill lead smelter] . . . has increased to the point where local action is inadequate and governmental action is needed to safeguard the health and safety of the employees and general population living in the area of the smelter."[59] The group went on to request an investigation into "the incidence of lead poisoning" at the smelter.[60] While company officials declined to respond to the allegations, NWMW president Eddie Adams said that his union was aware of the fact that "leading" had become more prevalent at the smelter, and contended that "we are fighting for an adequate solution to this problem."[61] To demonstrate that his union was working effectively to solve the problem Adams stated, "We have registered many complaints and have a meeting set up [with Bunker Hill management] to iron out the time of exposure these people have to gas, dust and fumes."[62]

While the dissidents and the Northwest Metal Workers Union evidently agreed on the nature and scope of the lead health problem at the smelter, if not on tactics to address the issue, the State of Idaho was dismissive. The state, represented by mine inspector George Fletcher, chalked the charges up to "inciters" involved in a petty intraunion squabble and declined to investigate conditions at the smelter, hazarding, "We know the dangers of lead poisoning and they are no greater than they have ever been."[63] The mine inspector's prompt refusal to investigate allegations of a lead poisoning problem at the smelter surely was gratifying to Bunker Hill officials, but it hardly would have been surprising. With the exception of the decision of the state legislature in the early 1930s to authorize a study of the condition of the waters of the Coeur d'Alene River system, and some air and water quality regulation in the 1970s, the State of Idaho consistently had maintained a tradition of inactivity regarding the environmental and health consequences of the activities of the state's most powerful mining company.

By the early 1960s Bunker Hill, contrary to its public façade, was aware that it faced a growing environmental and employee health problem. The chief difficulty issued from the fact that the company's zinc and lead plants, built in 1928 and

1917, respectively, were simultaneously showing signs of age and being forced into increased production. This circumstance invariably led to greater levels of pollution exposure for both plant workers and the surrounding local community. At the same time, area residents and Bunker Hill workers—members of a 1950s–1960s "consumer's republic" that increasingly valued environmental commodities such as outdoor recreation and clean air and water—were becoming progressively less tolerant of the air quality conditions within and outside Bunker Hill's processing plants.[64] As A. Y. Bethune, Bunker's manager of metallurgy, reported to the company's board of directors meeting in 1961, the public was "becoming more and more aware of, and critical of, our physical operation. No matter how we may think or whatever the facts may dictate, we are in for a harder and harder fight to control dust and fume."[65] Perhaps in response to shifting local attitudes, the company's proposed capital budget for 1960 advocated spending some of the roughly $300,000 that had been appropriated but not spent between 1957 and 1959 for pollution control at the smelter.[66] An expenditure of $43,000 was recommended for "dust areas carrying high lead content and with a high incidence of plumbism" and another $24,000 was to be spent to "correct a very dusty situation."[67]

Weber Visits

Notwithstanding Bunker's corrective measures, industrial hygiene engineer Clarence Weber's examination of the smelter in 1964 revealed a number of problems for the lead health of workers. Weber noted that "as most lead smelter dusts and fumes contain lead[,] maintenance workers lead exposures can be very high while performing some . . . jobs such as taking out or repairing old timber structures, building maintenance repairs and etc."[68] The industrial hygienist also observed that some smelter workers evidently needed to wear protective breathing apparatus in order to safeguard themselves from the environmental hazards posed by their workplace, writing, "At times respiratory protection is essential for these employees [maintenance workers] to maintain a safe status in regard to their lead health. Better housekeeping and operational dust and fume control will improve their environmental lead exposure but it cannot eliminate the lead health problem for these employees."[69] Weber concluded that "employees who can't wear a respirator should not be working at the lead smelter."[70] While noting that "the severity and incident rate of lead absorption have both lowered" at the smelter in recent years, Weber saw a number of areas in which there remained room for improvement.[71] In addition to his advocacy for an improved and enhanced use of respirators by at-risk workers, the hygienist also called for better efforts by workers to clean up their work areas and a more serious worker attitude toward the problem of plumbism.

Weber additionally suggested that overtime be kept to a minimum, "especially for employees with a lead health problem," and that workers who tested high for lead in urinalysis be immediately moved to jobs offering far lower lead exposure, rather than awaiting the results of doctor visits before moving leaded employees.[72]

A New Era Dawns

While Bethune's comments in 1961 were prescient, it is doubtful that he or any Bunker Hill managers grasped the extent to which the scope and pace of change concerning the environment and worker health would accelerate in the late 1960s and early 1970s. Whereas for decades Jack Bradley's classic formulation of a tradeoff of prosperity for pollution had held sway in communities across the United States, in the sixties and seventies such "bargains" became increasingly unpalatable. The same can be said with regard to regulation. While for its first eighty years Bunker Hill experienced little to no federal or state health or environmental control, in the late 1960s and 1970s an avalanche of federal legislation, much of it containing requirements for state enforcement, hit the company. Bunker suddenly found itself in a new and strangely challenging world of public opinion and federal and state regulation. Among the most demanding of the new federal laws that the company would face was the Occupational Safety and Health Act of 1970. The act, signed into law by President Richard M. Nixon in December 1970, set the lofty goal of ensuring all American workers "safe and healthful working conditions." To realize this new federal guarantee, the act created two new agencies, the Occupational Safety and Health Administration (OSHA), designed to both set and enforce workplace health and safety standards, and the National Institute of Occupational Safety and Health (NIOSH), a research institute housed within the Centers for Disease Control (CDC). The act granted OSHA the authority to investigate the premises of nearly all US workplaces within eight hours of having made a formal request to do so, a power OSHA would use frequently in the Coeur d'Alenes during the 1970s and early 1980s.

OSHA Arrives

OSHA first investigated the Bunker Hill lead and zinc plants in 1972. What the federal investigators found convinced them of the need for return visits to Kellogg. For starters, airborne lead concentrations at the smelter were well above OSHA's workplace standards. The agency and the company quickly agreed to a program to bring the smelter's ambient lead levels down to federal standards. Bunker would undertake twelve maintenance and construction projects at the facility, to be completed by April 1975.[73] Until that time smelter workers were supposed to wear respirators for health protection.[74] Shortly prior to OSHA's first visit to the Bunker Hill smelter, a NIOSH evaluation at the plant determined that 44 percent of the 190 smelter workers tested had abnormal concentrations of lead or cadmium in their bloodstreams. Approximately one-quarter of the tested workers reported having received at least one course of chelation therapy, the standard medical treatment for severe lead intoxication.[75]

A front-page editorial in a March 1971 issue of the NWMW publication, the *Carrier*, indicated the continuing relevance of plumbism for Bunker Hill workers. The piece charged Bunker with breaching its contract with the union in various ways, including failure to provide workers with "safe, sanitary and healthy work places."[76] Addressing potential naysayers, the anonymous author cited "the number of lead cases and respiratory ailments caused by acid fumes, SO_2 [sulfur dioxide] and silica dust at the Bunker Hill."[77] The editorial concluded with an allegation that many Bunker workers died "long before their time." The columnist ascribed this sad fact to "the grim reaper" and "industrial disease," and challenged, "A comparable age group from industries with better kept up and ventilated work places live longer and healthier lives."[78]

In the same issue of the *Carrier*, under the headline "A Tickled Man," the publication expressed delight at the fact that company foremen and supervisors, due to a strike by some local AFL-CIO craft unions, were having to perform dirty work in the smelter, demanding, "Tell me how you like working in the filth and conditions you've been putting your men into. See how you like getting your body covered with lead dust; see how you like wearing inefficient respirators; see how you like packing some lead around *in* your body. . . . If you're lucky you may get to enjoy some I.V.'s [for chelation therapy]." Addressing the issue of company punishment of workers who complained about unhealthy working conditions at the smelter, the writer was rueful. "My only wish is this—If you should complain, maybe the Company will be as fair to you as it has been to the employees it's discriminated against for trying to make it a better place to work." The piece concluded with an admonition for managers who temporarily were having to do dirty smelter jobs, to "take good notice and let it penetrate. Maybe, just maybe you'll learn something. Maybe our gripes haven't been so petty after all."[79]

Likely more as a response to regulatory pressures than to worker concerns, in the first half of the 1970s Bunker significantly increased the number of urine tests it performed on the roughly 500 smelter employees—from 2,736 in 1970 to 3,694 in 1974.[80] Disconcertingly however, the number of seriously leaded workers increased at an even more rapid pace. Blood tests and doctor visits, both triggered by high urine lead readings, went up from 300 to 570 and 242 to 470, respectively, between 1970 and 1974.[81] While in 1970 fifty-two leaded workers required medical action, by 1974 eighty-two did. Transfers and temporary layoffs of smelter workers due to lead health problems also increased during the first half of the 1970s, with transfers rising from twenty-two to twenty-six and layoffs from six to twenty-six.[82] By 1974 seventy workers were receiving intravenous treatments of the drug EDTA (i.e., chelation therapy) to reduce the amount of lead in their bodies. For ten of these workers, the treatment represented a repeated effort.[83]

In early 1974, the barrier between the work environment inside the smelter and the local environment of the Kellogg area was breached, a few months after the now infamous baghouse fire in September 1973. For the first months following the destruction of the smelter's air filtration system, smelter workers took the brunt of the heightened air lead burden inside the smelter, but in January Bunker management heard their complaints about unbearable working conditions and began venting the toxic effluent directly out the smelter's main stack.

Respirators and Blood Tests

In early 1975, in response to OSHA requirements, Bunker implemented new health safety standards at the smelter designed to lower blood lead levels. While the company stated that it was making significant efforts to clean up the interior of the smelter—by, among other things, hooding the blast furnaces, making modifications at the sinter plant, installing large new scrubbers, and shutting down for four days the previous fall for general cleanup and repair—Bunker Hill president James Halley said that these measures alone would not be sufficient in the short run to guarantee worker health.[84] The new company policy was that, in certain high-lead areas of the smelter and zinc plant, and for specific high-lead jobs, workers now would be strictly required to wear respirators at all times. Workers also were "encouraged" to wear respirators at all times in the smelter.[85] Under the new rules, all workers in "high-lead areas" also were required to trim facial hair so as to ensure a proper seal of the respirators, regularly check their respirators to make sure of proper functioning, refrain from eating, smoking, or carrying food or tobacco in open containers except in the lunchroom, wash hands and faces well before consuming food or tobacco, have work clothes washed daily by the company's laundry service, shower and change clothes before leaving the plant after work, and report as scheduled for blood lead testing.[86] While smelter workers did not object to most of these rules, many strenuously opposed the requirement that they wear respirators at all times, complaining that they were bulky, hot, and uncomfortable. Ken Flatt, a former smelterman who served as president of Local 7854 of the United Steelworkers of America (USWA) in the late 1970s and early 1980s, remembers that most workers in high-lead areas understood the need for respirators, but also recalls the devices' chief drawback, noting, "They're uncomfortable . . . they sweat."[87]

A final new company rule required that smelter and zinc plant workers keep their blood lead levels below 80 μg/dl, the OSHA standard for worker blood lead, or risk firing. Regardless of whether or not a worker was implicated in other policy violations, Bunker maintained that a blood lead level of 80 μg/dl or higher constituted prima facie evidence of "habitual violation" of the company's lead safety policies.[88] If after ninety days of being tested at 80 μg/dl or higher, a worker's blood lead had not dropped below that level, he would be fired.[89] Smelter and zinc plant workers—organized since April 1971 under the aegis of Local 7854—voiced strong objections to this provision of the company's new lead health rules. David Duthie, Local 7854 president, complained that the rule "puts the monkey on the employees' back. . . . It says that if you're leaded, it's your fault."[90] Halley responded that the policy was necessary to ensure that workers wear their respirators.[91]

Although its long-term plans may have included cleanup of the operations of its lead and zinc plants, in the short run Bunker Hill chose to rely upon its traditional policy of shifting the burden for worker health to the workers themselves. While in 1918 that policy rested upon such hygienic techniques as showering, handkerchief wearing, and avoiding constipation, by 1975 it relied almost entirely upon the use of respirators. Rather than reform the smelter's dusty, heavily leaded environment, Bunker focused on ensuring that workers in that environment employed specific methods and technologies to shield their bodies from harm.

Sterilization or Smelter Work

Local 7854 launched two formal grievances over Bunker's new lead health policies for smelter and zinc plant workers in 1975. The first contested the rule permitting the dismissal of employees with high blood leads. The second involved a rule intended to protect female smelter and zinc plant workers and their fetuses from the dangers of high levels of lead absorption. In an effort to comply with equal opportunity laws, in 1972 Bunker hired Judy Etherton, the first woman employed at the smelter since World War II. Between 1972 and 1976, Bunker Hill hired forty-five women for production jobs, thirty at the smelter and fifteen at the zinc plant.[92] In early 1975, the company announced that fertile women no longer would be allowed to work in high-lead areas of the smelter and zinc plant because lead had been linked to high rates of stillborn children and miscarriages in female smelter workers.[93] Unless the female workers could prove that they were too old to bear children or had been sterilized, they were transferred to jobs involving less exposure to lead. For the workers, the new jobs often were problematic. They tended to offer lower pay along with work that was less interesting or significantly more demanding physically than what they had been doing. Sheri Rutherford, the twenty-three-year-old former smelter worker on whose behalf Local 7854 filed its grievance, was required to stack sixty-pound zinc slabs in her new job at the zinc plant. When Rutherford proved unable to handle the demands of the job, Bunker fired her.[94]

In September 1980, OSHA fined Bunker Hill $10,000 for the company's "sterilization policy," the maximum penalty it could assess for a single violation.[95] The agency labeled the company's policy a "willful violation" and stated that Bunker, like other employers, did not have the right to make female employee choose between working in an unhealthy environment and sterilization.[96] Directly challenging the company's fundamental approach to the issue of worker health, OSHA spokesman James Foster said, "OSHA's basic philosophy is that the hazard should be eliminated, rather than doing something [like requiring sterilization] or appending something [like personal protection devices] to the worker."[97]

A number of female smelter workers, however, evidently chose continued employment with the company over the possibility of having children. OSHA reported that "a number of women" at Bunker Hill underwent sterilizations, though it could not provide a precise figure.[98] Although the "sterilization episode" undoubtedly smacks of discriminatory paternalism on Bunker's part—in a rather literal sense, for the rule truly represented an effort to protect not female workers but their prospective fetuses—it also represents a rare instance in which the company made a significant effort to protect human health at its plants.[99] While the choice the company offered female smelter and zinc plant employees clearly placed those workers in a difficult and unfair bind, it is somewhat ironic that Bunker received one of its largest OSHA fines for actions taken in an apparently genuine, if misguided, effort to safeguard health.

A Health Study

In response to the requirement that all workers in high-lead areas of the smelter and zinc plant wear respirators, the Steelworkers local complained to NIOSH that Bunker was not adequately protecting employee health. Local 7854 president Duthie claimed that the union had difficulty getting access to company records of the urine and blood testing programs, with the result that "we just don't know how widespread this problem [leading of workers] is."[100] Bunker Hill, in response, downplayed the need for a health study. Bunker president Halley opined that "you can work in that lead plant and follow the rules and have no problem whatsoever, unless you are one of those people who are extremely sensitive to lead," adding, "My two sons have both worked in the smelter during summers and I had no particular qualms about sending them in there."[101] Halley's lack of qualms about his sons' lead exposure notwithstanding, however, NIOSH went ahead and prepared a large and intensive study of the health of Bunker's lead and zinc plant workers and of working conditions at the plants. Conducted between 1975 and 1976, the study's results were released in 1985.

In October 1975, OSHA proposed substantial downward revisions of the air lead standards for places of work and of blood lead levels for those who worked with "the useful metal," calling for a reduction of the air lead standard from 200 to 100 μg/m^3 and in blood lead from 80 to 60 μg/dl. Bunker backed the Lead Industries Association's (LIA) efforts to oppose adoption of the agency's proposed air lead standard, and contended that the cost of complying with the new standards would be $17 million.[102]

Winds of Change

In January 1977 a new Democratic administration, the first in eight years, took the reins of power in Washington. Under President Jimmy Carter, OSHA became a more aggressive advocate of worker health and safety. More than any other figure, Eula Bingham, OSHA's undersecretary, personified the agency's new orientation. An industrial toxicologist who had worked at the Kettering Institute and the University of Cincinnati, Bingham was sympathetic to organized labor's health concerns.[103] In 1977, OSHA held public hearings on the workplace lead standards the agency had proposed eighteen months earlier. The hearings generated tens of thousands of pages of testimony and written comment from groups including labor advocates, industry representatives, feminists, and scientists.[104]

Much of the testimony was reminiscent of the ecological concerns raised in the 1950s and 1960s regarding the health effects of relatively low levels of toxins such as radiation and pesticides.[105] A new, more ecological approach to worker health had begun to develop during the 1960s and it represented a direct challenge to the older industrial hygiene model. Whereas the new approach often looked for subclinical manifestations of harm from industrial toxins, industrial hygiene sought to determine the highest threshold values of exposure to a toxin at which workers did not manifest overt symptoms of illness. Industrial hygienists and toxicologists, whose research generally was funded by industry, had developed workplace blood-level thresholds for various toxins. Dr. Robert Kehoe, for instance, who dominated the field of lead toxicology from the 1920s to the 1960s, was instrumental in developing the 80 μg/dl blood lead standard for workers that governed for decades.[106] Kehoe, whose work focused largely on the effects of lead on adult males, performed many studies that sought to prove that humans were perfectly safe at this level of blood lead. Ecologically oriented scientists who challenged the work of Kehoe and other industrial toxicologists, however, were concerned not just with bodily markers of toxins such as blood lead levels, but also with levels of environmental exposure. They pushed to require measurements of lead in the air in lead-producing facilities in addition to the traditional blood lead readings.[107]

The new approach to occupational health was reflected in the workplace lead standards that OSHA issued in November 1978. The standards, more ambitious even than those proposed by the agency three years earlier, phased in dramatic reductions in permissible blood and air lead levels over the coming decade. By 1989, air lead levels were to drop from 200 to 50 $\mu g/m^3$ and worker blood lead from 80 to 50 µg/dl.[108] In a victory for workers, the new standards also mandated that employees with blood lead levels above the maximum allowable amount be retained at their present levels of pay and seniority and transferred to lead-free jobs until their blood lead content returned to acceptable levels.[109] Bunker once again joined the LIA in a legal challenge to the new OSHA standards. In its suit the LIA argued that OSHA's standard was politically rather than technically based and that the agency had ignored its mandate to factor in economic impacts on industry when crafting the new rules.

Although in the first year of the new standard employees were to be removed from the workplace only when their blood leads reached the long-standing trigger of 80 µg/dl, Bunker still experienced difficulties with compliance and had to remove thirty-three workers from the smelter.[110] At the time roughly half of the company's smelter workers were experiencing lead levels of over 60 µg/dl and forty-seven registered above 70 µg/dl.[111] Facing the imminent need to remove a number of additional smelter workers when the federal standard dropped to 70 µg/dl on March 1, 1980, the company filed a motion to stay implementation of the standard.[112]

An Unhealthy Environment

Bunker Hill, as it happens, had ample reason to fear OSHA's increasingly stringent workplace lead standards. In February 1980, Michael J. Wright, a young industrial hygienist working for the United Steelworkers of America, toured the smelter. Wright found that lead dust levels in the dirtiest areas of the smelter—not coincidentally the same areas where nearly all of those removed for high blood lead levels over the previous year worked—"almost uniformly exceed[ed]" the 200 $\mu g/m^3$ air lead standard.[113] Wright discovered that there existed a geography of pollution within the smelter. Thirty-one of the thirty-three workers removed in 1979 labored in one of four areas within the facility: eight worked in the high-line area, where the mean air lead level measured a whopping 2,059 μg/m^3; seven in the lead refinery, with a mean of 540 μg/m^3; five in the blast furnace, where the mean measured 1,359 μg/m^3, and four in the silver refinery, with a mean of 440 μg/m^3.[114] Addressing Bunker Hill's reliance upon respirators for the protection of worker health at the smelter, Wright wrote, "No practical respirator program is capable of adequately coping with the excessive concentrations of lead dust . . . that exist in the Bunker Hill lead smelter."[115]

The industrial hygienist found "extremely poor housekeeping" to be the principal reason for the excessive concentrations of lead dust at the smelter, noting the tendency of leaded dust left in the workplace to continually recirculate, and thus to repeatedly poison workers. Wright added, "It would be a gross understatement to say that there are 'accumulations' of lead-bearing dust on the surfaces. . . . Every floor I observed was dusty. . . . In some areas of the smelter . . . there are significant accumulations of lead-bearing dust on every surface." The hygienist saw "standing piles of lead-bearing dust, which, in some cases, reached the depth of six inches to one foot," mentioning that piles of this size were "so egregious that it could independently constitute a citable OSHA tripping hazard." Wright even observed the presence of "lead-bearing dust" in the smelter's lunch and shower rooms, an area where workers necessarily did not wear respirators. While noting that Bunker had "equipped many of these facilities with positive pressure, filtered air systems," the hygienist also pointed out a number of "fundamental deficiencies"—namely, inadequate clothing decontamination and lunchroom door seal systems and the lack of any procedure for the rinsing of lead-bearing dust from boots before entering the facilities.[116] While piping clean air into shower and lunch rooms was helpful, on its own this measure hardly rendered spaces lead-free when dusty clothes and boots still were permitted to enter.

Wright also found the smelter to be seriously deficient in dust-control technology, faulting Bunker for failing to implement "numerous control devices that have been available for quite some time."[117] Among other things, the hygienist recommended that the company enclose and ventilate conveyers used to move lead sinter and lead ore, place a portable hood over a drossing operation where he observed the creation of "a thick cloud of fumes and lead-bearing dust," employ better hooding and local exhaust ventilation on the blast furnace tapping operation, and enclose the cabs of the numerous front-end loaders, tow motors, forklifts, and other vehicles used in the plant to protect their drivers from lead fumes. Wright cited these improvements as examples of the sorts of technological improvements that a company serious about complying with the nine-year-old workplace air lead standard of 200 μg/m^3 would have long since made.[118]

Comparing the situation at the Bunker Hill smelter with what he had seen at the Ronnskar lead smelter on a recent visit to Sweden, Wright noted that while air lead levels at Ronnskar were roughly 200 μg/m^3, during the previous year only one worker there had registered a blood lead level of over 70 μg/dl. Wright attributed Ronnskar's worker lead-contamination record, far superior to Bunker's, primarily to the Swedish company's intense focus upon "rigorous housekeeping," along with "clean locker, shower and control rooms."[119] Though normally in cases where respirator use programs were in force worker blood lead levels would be expected to decline, Wright judged that given the Bunker Hill smelter's "inordinately excessive concentrations of lead dust," a rise in the number of leaded workers was possible.[120] In the industrial hygienist's view, the air and blood lead problems at the smelter were the direct results of conscious choices made by the company.[121] It thus was fully within Bunker's power to effectively address its environmental lead and worker health issues at the smelter. The company simply needed to be willing to install new pollution-control technologies and improve its housekeeping practices.

OSHA Draws Fire

In 1979, OSHA came down heavily on Bunker Hill, levying over $50,000 in fines against the company. Early in the year, the agency fined the lead producer $13,700 for safety violations at the smelter. In May, in the largest fine the agency had yet levied against an Idaho employer, OSHA added nearly $37,000 to the company's tab.[122] While still tasking Bunker for permitting excessive worker exposure to lead at the smelter and zinc refinery, the agency added a new charge—allowing workers to be exposed to too much arsenic. While OSHA recently had lowered its standard of employee exposure to arsenic by a factor of fifty, from 500 to 10 μg/m^3, the company was penalized for violations of the old standard. In all Bunker was charged with more than thirty-five violations. These included the failure to provide all smelter workers with respirators, letting workers consume food and beverages in more than a dozen arsenic-laden lunchrooms, failing to warn workers of the presence of potentially carcinogenic compounds in the lead and zinc plants, refusal to turn over employee health records to federal safety inspectors, and failure to provide separate storage areas for workers' work and street clothes.[123]

Following its by-then-routine game plan, Bunker immediately contested OSHA's charges and the accompanying fines. The company already was on record as opposing the arsenic standard as "unworkable" and "not supported by any reasonable evidence of adverse health effect."[124] By the late 1970s Gulf-Bunker's frustrations with the new regulatory regime were showing. The company had developed a largely antagonistic attitude toward federal health and environmental agencies, with company officials routinely taking potshots at OSHA and the EPA in the press. Following the May 1979 fines, Bunker's vice-president for environmental affairs, Jim Boyd, sniped, "OSHA has not been making working conditions safer. There are more accidents now than before."[125] In a June 1978 speech in Kellogg marking the tenth anniversary of Gulf's acquisition of Bunker Hill, Gulf CEO Robert Allen blamed federal regulatory agencies for bringing major natural resource companies "to the brink of economic disaster."[126] Painting EPA and OSHA officials as part of an "arrogant bureaucracy," whose overzealous actions made him "gravely concerned for the future of Bunker Hill," Allen singled out OSHA for adopting standards "that are clearly as impossible to meet as they are unnecessary."[127]

In late May 1979, a visit to Kellogg by Idaho Republican US Senator James McClure fanned the flames of a small dissident worker protest against OSHA. McClure, a second-term senator who enjoyed unusual popularity in heavily Democratic Shoshone County, was known as a friend of the mining industry. During his stop in Kellogg, McClure visited with a number of Bunker Hill lead and zinc plant workers, some of whom had asked him to hear their concerns about recent OSHA requirements.[128] At the meeting, workers complained to the senator about the "ill-fitting" coveralls and "bulky" respirators they had been required to wear in the wake of OSHA's latest round of sanctions and stated that some of their fellow employees were resigning rather than comply with the new equipment requirements.[129] McClure exacerbated workers' concerns at the meeting by suggesting that excessive OSHA regulations ultimately could lead to the closure of Bunker Hill's operations.[130] A few days after the senator's visit, the group of dissidents sent a letter to all their fellow Local 7854 members. Referring to OSHA as "a Frankenstein monster" that had been created with union support, the letter charged that the agency was threatening their jobs and contended that despite strong support from both the Kellogg local and the national union (USWA), OSHA remained "the federal agency that is making regulations that make our jobs ever more unbearable."[131]

In early June Senator McClure stated that OSHA's new arsenic exposure standard was "threatening the existence" of jobs at Bunker Hill, called the standard "excessive," and said it was "hampering" the work of Bunker Hill employees. The Idaho senator also requested a formal examination of the arsenic rule. Shortly thereafter Local 7854 held a meeting to air grievances over the new protective gear requirements. The meeting drew approximately two hundred workers. There were complaints over the forced use of respirators at the lead and zinc plants, but no groundswell of opposition to OSHA materialized. Countering Bunker's charge that OSHA regulations were forcing the company to mandate the wearing of coveralls and respirators in all areas of both the lead and zinc facilities, OSHA's Idaho director Richard Jackson told workers that in fact Bunker was requiring use of the protective gear in areas of the plants where the agency did not feel it was necessary. Local 7854 president Ken Flatt stated that in plant lunchrooms and dressing rooms the gear became necessary because the company had refused to end the lead and arsenic contamination of those facilities. Following the meeting, an obviously frustrated Flatt accused Senator McClure of "playing politics" with the issue of worker safety.[132] While it is clear that at this time some smelter and zinc refinery workers were willing to side with conservative politicians and Bunker Hill managers in their criticisms of OSHA, the rank and file remained unwilling to join in their protests.

Far from backing down in the face of attack from Bunker and its political allies, in September 1980 OSHA cited the company for 108 violations and assessed it $82,765 in penalties. While the fine for Bunker's "sterilization policy" garnered much of the public's attention, the company faced a litany of other workplace safety violations, many of them repeat offenses. Despite high earnings due to record-high silver prices the previous year, Gulf-Bunker evidently still had not cleaned up its lunchrooms—the company was fined $10,000 for "excess lead exposure in eating areas." The company also was cited for repeat violations for, among others, failure to monitor workers' blood leads, failures in record keeping, and failure to ensure adequate use of respirators by lead and zinc plant workers. Not surprisingly, the company also was rapped for its failure to comply with the new arsenic standard.[133]

Shifting Winds Prompt a Different Tack

In August 1981, Gulf Resources stunned northern Idaho with the announcement that it planned to shutter its Bunker Hill operation by the end of the year unless a buyer for the ninety-four-year-old company could be found. It soon came to light that the Texas company had shopped its Idaho subsidiary for two years with no luck before making the announcement. The prospect of Bunker Hill's closure—devastating to the economy of north Idaho where the company's 2,100 jobs and annual payroll of $50 million represented sizable fractions of the whole—quickly galvanized people to action.[134] In the near-frantic efforts to entice a buyer for Bunker's aging mine and plants that ensued, Local 7854 proved more than willing to make major concessions. The union local offered a 15 percent pay cut and worked with OSHA to effect less onerous lead health compliance requirements. As Flatt succinctly summarized his union's efforts to roll back health regulations it had worked for years to put in place, "When you start playing with a guy's livelihood .

. . things change."[135] With strong bipartisan support from Idaho's Democratic governor John Evans and the state's Republican congressional delegation, Local 7854 pressured OSHA into granting a prospective Bunker Hill purchaser a five-year exemption from the requirement that workers be removed from the smelter and zinc plant when their blood lead levels reached 60 μg/dl.[136] Agreeing to this deal on behalf of the new Reagan administration was OSHA head Thomas Auchter, who stated that while protecting workers "is our paramount concern," the need "to sustain jobs for workers can't be ignored."[137] As part of the deal, however, Gulf-Bunker made some concessions to Local 7854 and OSHA, agreeing to spend $880,000 to finally clean up lunch and break rooms at the smelter and zinc plant, to settle an OSHA complaint from the previous year, and to increase employee medical checkups from once to twice a year.[138]

In the end, however, all of the efforts to find a buyer for Bunker came to naught and the last of the company's works closed in February 1982. The desperate struggle of the last final months to make the company salable, though, shed some revealing light on the core values of workers, community members, and Idaho politicos of various ideological stripes. When it became clear to all parties that they were faced with the imminent shutdown of the largest business in northern Idaho, normally antagonistic interests such as labor and management and Republicans and Democrats came together to try to save the company. In addition, to elaborate upon Flatt's blunt assessment, issues like blood lead and long-range health concerns suddenly became far lower priorities to workers who were faced with the impending loss of their employment. Jobs in Bunker's lead and zinc plants, while undeniably dirty and often unhealthy, were relatively high-paying. Many who labored at the zinc and lead plants were older workers who had held relatively specialized, low-skill jobs for years. For such workers the likelihood of finding new employment at comparable wages was quite low. These factors undoubtedly compounded the workers' desires to do whatever they could to preserve their jobs.

Until the advent of the "environmental era," circa 1970, with but a few exceptions Bunker's lead and zinc plant workers had been willing to use health concerns as a bargaining chip with which to gain higher wages or greater benefits. Following the creation of OSHA, however, the workers, now with the federal government in their corner, began an outright push for worker health studies and improved health and safety conditions at the plants. It must be understood that Local 7854's concessions on health safety in 1981, while notable, represented merely a temporary stratagem intended first to preserve jobs by attracting a buyer for Bunker's aging smelter and zinc refinery and then later to help that purchaser to "get on its feet." Once the presumptive goal was reached—a five-year grace period was offered on

OSHA blood lead standards—OSHA's lead and other health standards were expected to kick back in at the plants. While Bunker's smelter and zinc plant workers were willing to make short-run health concessions to preserve their jobs, they were not willing to work in unhealthy plants once their presumptive new employer's survival was secured.

Health and Mortality Studies Offer Insight

At the request of Local 7854, in 1975 and 1976 NIOSH, in conjunction with the Bureau of Epidemiology and the CDC, conducted an industrial hygiene study of Bunker Hill's lead smelter and zinc refinery, along with medical surveys of workers at those plants. The purpose of the investigation was to determine both the health of the environment inside the smelter and refinery and also the epidemiology and health consequences to workers of exposure to heavy metals. The study examined 173 male and 48 female smelter workers and 209 male zinc refinery workers.[139] For purposes of comparison, the study also examined 97 male and 42 female Kellogg residents who did not work at the smelter or zinc plant, and a presumably "nonleaded" control group of residents of Spirit Lake, a community forty-five miles from Kellogg.[140]

The industrial hygiene portion of the investigation revealed air lead concentrations at the smelter that were even higher than those Michael Wright would find five years later. Mean air lead exposure measured 3,530 μg/m^3, more than seventeen times the prevailing OSHA standard of 200 μg/m^3 and more than 50 percent higher than Wright would find in the area of highest concentration during his visit.[141] Smelter workers also were exposed to high levels of zinc and arsenic.[142] At the zinc plant mean lead levels were lower, measuring 401 μg/m^3, but that was still twice the recommended maximum. Zinc plant workers also were exposed to high mean concentrations of zinc (6,850 μg/m^3), cadmium (38 μg/m^3), and sulfur dioxide (6.72 ppm). Arsenic levels, at 7 μg/m^3, although lower than those at the smelter, were above the proposed OSHA standard. The unfortunate workers in the zinc plant's baghouse, however, were subject to a stratospheric mean lead exposure of 5,590 μg/m^3.[143] Predictably, the study also found that worker blood lead essentially mirrored the environmental levels found in the facilities where they worked. Nearly two in five smelter workers tested (38 percent) had blood lead levels of 60 μg/dl or higher, while the same was true for 14 percent of tested zinc plant workers.[144]

When the NIOSH researchers turned to the effects of such elevated concentrations of lead and other heavy metals on the smelter and zinc refinery workers, they found a number of maladies. Their report's official summary stated that "gastrointestinal, constitutional, and neurological symptoms were significantly more prevalent than in community residents. Also, the frequency of those symptom complexes correlated positively with blood lead . . . concentrations." NIOSH also discovered that the lead smelter workers "demonstrated an increased prevalence of symptoms of peripheral nerve dysfunction as compared to community residents."[145] In addition, the study revealed that the smelter workers manifested increased feelings of depression and hostility, lengthened choice reaction times and a significant decrease in motor nerve conduction velocity (1–3 m/sec).[146] Perhaps even more discomfiting for Bunker's smelter employees was NIOSH's finding of "an increased prevalence of tubular renal dysfunction and of disordered calcium metabolism in Bunker Hill workers."[147] Neither finding was particularly surprising, however, as high levels of lead ingestion was strongly correlated with kidney problems and lead was known to substitute for calcium in the body.

As compared with Kellogg community residents, hardly a lead-free population, lead and zinc plant workers were found to have elevated urinary protein and blood serum calcium levels. Evidencing the heightened health risks of working in certain high-lead areas of the plants, abnormalities in blood calcium and phosphorus concentrations, along with elevated B-2-microglobulin levels, "were found in many male workers in several work areas in both the lead smelter and zinc plant." Abnormalities in renal function and elevated uric acid levels also were found in workers who had been treated by chelating agents for lead intoxication. The severity of these problems increased with the number of courses of chelation therapy. Far from being "cured" of lead health problems once chelation therapy had lowered their blood leads enough that they could return to work, the evidence indicated that lead-intoxicated workers often became trapped in a vicious cycle that made them more susceptible to serious health problems.[148]

A Mortality Study

An article on the mortality rates and causes of death of workers at the Bunker Hill lead smelter appeared in the *American Journal of Epidemiology* in 1985. The article was an outgrowth of the NIOSH study of 1975–1976. Philip Landrigan, Bunker's longtime antagonist, served as one of the article's four coauthors. The NIOSH study upon which the article was based examined mortality rates for 1,987 men who worked at the Bunker Hill smelter for at least one year between 1940 and 1965.[149]

The study found that as of December 31, 1977, 1,281 workers (64.5 percent) were still living, 665 (33.4 percent) were deceased, and 41 (2.1 percent) could not be accounted for.[150] Although the investigation found that the smelter workers' mortality rate was roughly equal to that of the United States' white male population as a whole, with a standardized mortality ratio (SMR) of 98, it also revealed excessive mortality rates from specific sources.[151] The study found significantly heightened death rates from chronic renal disease in long-term lead workers (SMR = 192) and an increase in mortality from renal cancer (SMR = 204). The renal disease finding was consistent with the results of four other studies of occupational lead exposure, while the renal cancer discovery was consistent only with experimental studies on animals.[152] For workers with more than twenty years of lead exposure, the SMR for chronic renal disease rose to 392—four times the US average for white males. Describing the sorts of kidney problems that beset many Bunker Hill smelter workers, the article's authors wrote, "Nephritis (inflammation of one or both kidneys) has been recognized as a component of occupational lead intoxication since the late nineteenth century . . . it appears that nephropathy (damage or disease to the kidney) is a consequence of prolonged, relatively high-dose exposure to lead. The end stage of neuropathy is chronic renal failure."[153]

Renal problems were the most serious, widespread health issue uncovered by the mortality study, but they were not alone. Investigators found that for those who worked longer at the smelter mortality rates were heightened, with "an upward trend over time in mortality from cardiovascular disease."[154] The data indicated that the longer one worked at the Bunker Hill smelter, the more likely it was that one would develop a serious and possibly fatal medical condition.

The researchers also found elevated mortality rates from emphysema, silicosis, and other occupational respiratory diseases. These results were largely discounted, however, due to the fact that a considerable percentage of the smelter workers also had worked as hard rock miners in the Coeur d'Alenes, where they experienced a reasonable likelihood of contracting a respiratory disease.[155] A high number of smelter workers also died from accidents, but this again seems to have been related to the common practice of moving back and forth between employment at the smelter and the area's numerous mines.[156]

Updated in 1985 and 1988, the NIOSH mortality study continued to show high rates of death for Bunker Hill smelter workers from kidney disease and kidney cancer, along with elevated death rates from a new source: stroke. By 1988, when nearly four hundred more workers had died, deaths from kidney disease remained at four times the national average for white males, while mortality from renal cancer ran at nearly double and stroke at one-and-a-half times the national averages.[157] The updated reports offered further evidence that working in the Bunker smelter was bad for workers' health in general, and that it took a special toll on their kidneys.

The dirtiness of the Bunker Hill smelter did not come as news to those who labored there, of course. Workers had long maintained that the smelter was overly dusty and dirty. For just as long, they had expressed concerns about the effects of that leaded dust and dirt on their health. Health inspectors, too, since Royd R. Sayers's visit in 1918, had called upon the company to take steps to clean up the facility. While the company had at times made efforts at improving hygiene at the smelter, and also occasionally had purchased new, cleaner-operating equipment, these efforts had not kept pace with industry standards or with the pollution problems inside the aging facility. By the 1970s, the over-fifty-year-old smelter, and the decade-younger zinc refinery, were far out of compliance with federal standards for ambient levels of air lead. In response, Bunker adopted a policy of mandatory respirator use at the smelter in a somewhat desperate attempt to comply with OSHA requirements for worker blood lead levels. As Wright's study in 1980 showed, however, despite Bunker's efforts to insulate smelter employees from their work environment, their blood lead levels still remained far above federal safety standards.

Throughout the existence of its lead smelter and zinc refinery, Bunker Hill remained largely wedded to a strategy of placing the onus of safeguarding worker health upon workers themselves. This strategy, consistent with the company's cost-savings ethos, supports Arthur McEvoy's observation that "just as they do in environmental areas, . . . market forces overdiscount the costs of industrial injuries (thus subsidizing employers at the expense of injured workers) particularly when the harms involved are long-term, hard to identify, or difficult to value in money terms."[158] In the 1920s, Bunker's worker-health strategy took the form of placing prominent warnings of the dangers of lead intoxication throughout the smelter and encouraging workers to practice personal hygiene. By the 1970s, the strategy found expression through respirator requirements and the frequent blood lead testing of workers. As the 1980s dawned, the company apparently remained unwilling to make the technological changes, or to embark on the hygiene campaign necessary to bring the smelter into compliance with current health regulations. By that point, however, Bunker's hygienic-regulatory problems at the old smelter had become more severe, and the company's fiscal position significantly weaker, than they had once been. Had Bunker instead chosen to overhaul or replace the facility in the mid-1960s, when the company was fiscally robust, it might have been able to weather the challenges of the new regulatory era of the 1970s.[159] By the 1970s the company, at that point a subsidiary of Gulf Resources, was increasingly beleaguered and unable to keep pace with the regulatory and competitive burdens it faced. In 1982, when Bunker finally succumbed to those burdens, one of the chief reasons for the company's failure was its dirty, outmoded smelter.

As is the norm under capitalism, the burdens and risks of the Bunker Hill smelter were not shared equally. Bunker's lead and zinc plant workers occupied distinct "social risk positions" within the company and the community.[160] These decidedly risky social positions were held by workers, many of them unskilled, who lacked class status. While the company's closure threw nearly all of its employees out of work, only Bunker's smelter workers (and to a lesser extent its zinc plant workers) were left to shoulder the significant health costs of years of working in environments that were saturated with lead and other heavy metals.[161] As the NIOSH health study of smelter and zinc plant workers in 1975–1976 and the agency's later mortality studies clearly showed, serious health issues, and in some cases premature death, went hand in hand with working in the Bunker Hill smelter and zinc refinery. Over time workers at these facilities "paid for their work" with a host of maladies that included anemia, slowed nerve conduction, colic, hypertension, depression, stroke, and various forms of kidney malfunction and disease. As Peter Piekarski, a twenty-seven-year veteran of the smelter, longtime union activist, and kidney disease sufferer, said following the release of the results of the NIOSH mortality study of 1988, "I could've told them those results a long time ago. . . . We knew we were working in a hazardous industry."[162]

CHAPTER 7

A BRAVE NEW WORLD

> With reactor accidents or chemical catastrophes, "blank spots" on the map arise again in the most advanced stage of civilization. They are monuments of what threatens us.
>
> **Ulrich Beck**

Stunned disbelief greeted the announcement. On August 25, 1981, Bunker Hill announced over Osburn radio station KWAL that it would be liquidated and its Kellogg works closed by year's end unless its parent company, Gulf Resources & Chemical Corporation, could find a buyer for the ninety-four-year-old company, producer of 20 percent of the nation's silver and 17 percent of its lead. The announcement triggered a frantic, roller-coaster search to find a buyer for "the Bunker," whose workforce of 2,100 and $50 million annual payroll made the company northern Idaho's most significant employer by a wide margin. Idaho's senior US senator, James McClure, telephoned Vice-President George H. W. Bush at Bush's vacation home in Kennebunkport, Maine, to enlist his support for imposing a five-year moratorium on the application of any new federal environmental regulations to the Kellogg concern.[1] North Idaho's Republican congressman, Larry Craig, in a meeting with President Ronald Reagan, briefed the President on the imminent closure of Bunker Hill and on the devastating impact that it would have on the north Idaho economy.[2] Idaho governor John Evans created a special committee, the Silver Valley Economic Task Force, to develop a formula to attract a buyer, and assigned aide George Tway to assist the Kellogg-based group.[3]

Assessing the significance of losing Bunker Hill, Tway remarked, "This would be one of the most serious economic disasters the state has ever seen."[4] A study by Eastern Washington University economics professor Shik Young supported Tway's grim assessment. According to Young, if jobless workers did not emigrate following the Bunker Hill closure, the unemployment rate for Shoshone County soon would approach 50 percent and neighboring Kootenai County's would hit 20 percent.[5] The economist concluded that the closure would cost Shoshone County roughly 3,570 jobs and $67.5 million in lost wages, while Kootenai County stood to lose 1,680 jobs and $27.5 million in wages. More distant Spokane County,

Washington, according to Young, would escape with less damage, suffering losses of "only" eight hundred jobs and $14.2 million in wages.[6]

Although Idaho's Democratic governor and its Republican congressional delegation pulled together in efforts to find a buyer for the mining and smelting giant, their task was daunting. Gulf president Robert Allen had announced projected losses of $40 million for the coming year for Bunker.[7] The severe headwinds facing the company included metals prices that were hitting rock bottom; an insufficient supply of concentrates to keep the company's lead smelter running at close to capacity; an ongoing recession that showed no sign of letup; Bunker's aging smelter, which faced an expensive makeover to comply with EPA and OSHA regulations; and the company's large, relatively high-cost, unionized workforce. Throughout the late summer and fall, Bunker proceeded with shutdown efforts at its industrial complex, laying off workers and shuttering operations. On December 11, however, Silver Valley residents were granted a reason for hope when it was announced that a group of investors that included Wallace mining magnate Harry F. Magnuson and Coeur d'Alene publisher-developer Duane Hagadone had acquired an exclusive purchase option "for the entirety of Bunker Hill's holdings."[8] The streets of Kellogg were quiet between eleven and noon as people crowded around radios to listen to KWAL's broadcast of the Coeur d'Alene news conference announcing the pending sale of the company.[9] The investor group, though prepared to spend $65 million for Bunker, demanded major concessions from the union representing most of Bunker's workers and from the federal government. From Steelworkers Local 7854, the investors called for, among other things, a 25 percent pay reduction for the coming year, the right to assign workers without regard to seniority, and a reduction of medical insurance payments.[10] With the assistance of Local 7854, the group successfully cajoled the federal government into allowing a five-year moratorium on the application of new environmental and work safety rules.[11] On January 11, 1982, the investor group was bolstered by the addition of J. R. Simplot, the Idaho "potato king," who emerged as the consortium's majority investor.[12]

Firm opposition to the deal from the union remained, however. Local 7854 president Ken Flatt urged his membership to reject the Simplot group's proposal, arguing that doing so would bring the investors back to the table to negotiate a better agreement. Hagadone, speaking for the investor group, averred that he was presenting its last, best offer.[13] On January 17, Local 7854 and Bunker's six small craft unions assembled to vote on the proposed contract. By a 695 to 506 (58–42 percent) margin, the workers voted to accept the deal. The vote was ruled to be merely "advisory," however, and thus not binding, by the leadership of the Pittsburgh-based International Steelworkers Union, representatives of which were on hand for the session.[14] The International's leadership evidently was greatly concerned that if the proposed contract were adopted, its reduction of employee retirement benefits would lead to numerous lawsuits from aggrieved Local 7854 members.[15] Following the invalidation of the vote by its parent organization, Local 7854 split into warring factions. While one group remained loyal to Flatt, a large splinter faction emerged, led by Ron Byrd.[16] The splinter group, which favored ratification of the proposed contract, enjoyed strong community support.[17] The investor group announced that, unless the International Steelworkers promptly approved the January 17 vote, its efforts to purchase Bunker were at an end.

This announcement swung Idaho politicians and Kellogg residents into another bout of frenzied action. Bunker Hill salaried personnel supporting ratification of the contract sent a petition protesting the Steelworkers position to Congressman Craig, who promised to deliver it to President Reagan. The petition contained the signatures of over 6,500 Silver Valley residents.[18] KWAL broadcast the telephone numbers of the executive offices for the White House, the vice-president, and the assistant secretary of labor, while Silver Valley citizens sent an estimated one thousand telegrams to President Reagan, petitioning him for help in rescuing Bunker Hill.[19] On February 2, Idaho senator Steve Symms and Congressman Craig met with International Steelworkers president Lloyd McBride in Pittsburgh in an ultimately unsuccessful effort to persuade him to sign the agreement that had been approved two weeks earlier by a majority of Bunker workers.[20] The International Steelworkers' stance created significant bitterness in the Silver Valley. Some likened it to the contemporaneous situation in Poland, where the Eastern Bloc government was cracking down on the prodemocracy trade union Solidarity. Others drew comparisons to the way Bunker was being treated by another out-of-state concern, Gulf Resources.[21] As it happened, the effort to persuade McBride represented a last gasp for the months-long effort to save Bunker Hill. Following its failure, the investor group withdrew from the field, and Gulf sold off much of the old Bunker works for scrap.[22]

The most recent phase in the history of Kellogg and the rest of the Silver Valley, the postindustrial period that runs from 1982 through the present, begins with the radical shift in economy that was forced upon the area with the closure of the Bunker Hill operations and the ensuing curtailment of the Valley's mining industry. Although mining has persisted in the Silver Valley, its character has been significantly more intermittent and less robust than it was before the 1980s. A few years after Bunker's shutdown, the citizens of Kellogg decided to reinvent their town as a Bavarian-themed skiing community. The fateful decision to become a "tourist town" has shaped much of the subsequent history of the former mining and smelting hub. Whether its embrace of the "devil's bargain" of tourism has been beneficial for Kellogg remains an open question.[23]

Deindustrialized Communities

The contemporary, postindustrial story of Kellogg and the rest of the Silver Valley has many parallels. Over the past forty-five years, numerous American communities have faced the closure of a principal industry as the nation has shifted from an industrial to a services and information economy. From steel cities such as Youngstown, Ohio, and Pueblo, Colorado, and copper enclaves like Butte and Anaconda, Montana, to textile towns such as Mount Airy, North Carolina, the story has been the same—long-standing and relatively high-paying blue-collar jobs have been lost, and communities have been forced to redefine themselves and find a new economic rationale. A number of these deindustrialized communities, like Kellogg, also suffer from the legacy of industrial pollution.[24] Lying at the heart of a very large Superfund site, Kellogg perforce has conducted its reinvention efforts in the midst of one of the EPA's most challenging cleanups.

One of the hallmarks of global capitalism is its dynamism. This aspect of capitalism is manifested in myriad ways—from the global proliferation of smartphones, and all the ready access to information and culture they represent, to the mushrooming of a new manufacturing plant virtually overnight almost anywhere in the world. Capitalism's dynamism, however, can just as easily betoken the loss of thousands of jobs that accompanies the sudden shutdown of an old factory, or an oil rig explosion that leads to worker deaths and the fouling of hundreds of square miles of seawater and beach. In the idiom of Marshall Berman (with a debt to Karl Marx), "to be modern . . . is to experience personal and social life as a maelstrom, and to find one's world and oneself in perpetual disintegration and renewal, trouble and contradiction: to be part of a universe in which all that is solid melts into the air."[25] The past thirty-five years of Kellogg history have granted residents of that community a particularly vivid experience of modernity as maelstrom.

A Difficult Day Dawns

The unraveling of the Simplot group's efforts to purchase Bunker Hill and keep it running plunged Kellogg residents back into precisely the situation they had been dreading since late August. By April 1, all 2,100 Bunker Hill employees would be released. Though nearly all faced challenges, those faced by the seven to eight hundred unskilled workers at the lead and zinc plants appeared to be the greatest. According to Bunker president Jack Kendrick, while skilled craft workers and miners might face a reasonable chance of finding work elsewhere under decent economic conditions, that was not the case for the "cadre of loyal and dedicated unskilled workers" who ran the company's surface facilities.[26] Flatt estimated that only 10 percent of former Bunker Hill workers would be able to find jobs.[27] While referring to himself as "in the lucky 10 percent" who had found work, former fertilizer plant superintendent Bob Smith regretted that his new position as a plant manager in Omaha, Nebraska, would force him to leave his wife, daughter, and home behind in Kellogg. Smith, a Kellogg native and third-generation Bunker Hill employee, voiced the shock that many locals must have felt over recent events and their implications for a long-established way of life in the Coeur d'Alenes, saying, "Bunker Hill was always there. Jobs were plentiful and the pay was good. . . . The kids in school could say, 'I don't want to go to school, I want to work for Bunker'. . . . The kids who wanted to go to college could go. The kids that wanted to work could do that also."[28]

To address the unemployment situation in and around Kellogg, federal officials appropriated $700,000 for job retraining and Governor Evans turned over $50,000 from the governor's emergency fund for job search training.[29] Senator McClure secured special trade adjustment assistance from the US Department of Labor that enabled about 1,200 former Bunker workers to receive up to twelve months of unemployment benefits.[30] A federal grant also made high school equivalency classes available to locals. In 1982, eight hundred Kellogg residents returned to the classroom. When "the Bunker" was active, a high school diploma was not needed to get a good job, and three out of four Silver Valley workers lacked one.[31]

The valley's nascent economic realities demonstrated the need for the federal and state aid. In response to a $600,000 drop in revenues, the Kellogg School District closed a school and eliminated thirty-four positions during the 1982–1983 school year.[32] The City of Kellogg's assessed property values dropped 44 percent between December 1981 and December 1982.[33] The unemployment rate for Shoshone County was 30 percent in May 1983.[34] By August of that year, the Kellogg School District had eliminated more than forty positions. Things would have been even worse for the district had it not received over $323,440 in emergency state assistance.[35] In a sure sign of the times, Wallace High School's program in mining education, unable to secure employment for its eighteen students at mining companies during the summer of 1982, instead sent them to work as guides for tourists visiting the nonworking Sierra Silver Mine.[36]

Census data further highlight the image of an area reeling from an economic implosion. Between March 1981 and March 1983, the number of employed persons in Shoshone County fell from 6,203 to 3,532, or about 43 percent, while annualized payrolls fell from $125,119,000 to $78,593,000 (about 37 percent).[37] The ripple effect of the gutting of the Silver Valley's industrial base can be seen through the impact on the county's service sector. Over the two-year period 24 jobs were shed in wholesale trade, 221 in retail trade, 44 at auto dealers and gas stations, 116 at dining establishments, and 149 at hostelries and recreational services establishments.[38] A number of the valley's service industry workers joined the former Bunker Hill miners and plant workers in looking for work in 1982–1983, their jobs having been eliminated after paychecks to local miners and plant workers disappeared.[39]

An indicator of the impact the Bunker Hill closure had on the regional (Inland Empire) economy is the effect it had on the Union Pacific Railroad. Bunker had been the Union Pacific's most important shipper in the Spokane region. Every week forty railcar loads of lead and zinc concentrates and phosphate rock had traveled from Spokane to Kellogg and twenty-five to thirty-five carloads had carried lead and zinc ingots, fertilizer, cadmium, and sulfuric acid on the return trip.[40] The Bunker Hill closure led to the furloughing of eight crews—thirty-two trainmen—who had worked on the daily trains between Spokane and Kellogg. By mid-1983 the railroad was sending only two trains a week on the route.[41]

Five years later, despite a rebound in the price of silver and the reopening of some area mines, the Silver Valley economy was still reeling.[42] In 1987 Shoshone County's unemployment hovered at around 50 percent. The county, long one of Idaho's most prosperous, had become its poorest. Lack of work drove many out. The county's population had dropped by one-quarter to roughly fifteen thousand.[43] Quite a few of the valley's mining families appear to have left the district during the disastrous year of 1982, searching for work in other western mining and smelting communities. According to Ray Frank, a former miner at the Sunshine Mine who in the mid-1980s was collecting statistics for the Idaho Department of Employment's Job Service Office in Kellogg, families had relocated to places such as Silverton, Colorado; Troy, Montana; Rock Springs, Wyoming; Trail, British Columbia; Republic, Washington; and El Paso, Texas.[44] Larger numbers of Silver Valley families, however, about fifty in each case, migrated to San Miguel, Arizona, and the Stillwater Valley in south-central Montana. San Miguel boasted a large copper mine and the Stillwater was the site of a new platinum mine. Many of the families still owned homes and had family back in the Coeur d'Alenes. While a number of the refugees expressed nostalgia for northern Idaho and for the higher wages they had been paid in the Silver Valley, no one seemed to miss the job insecurity that had become part and parcel of working in the Coeur d'Alenes.[45] Other paths were taken, too. According to Frank, about eighty-five miners stayed in Kellogg to seek retraining, while an undetermined number left in 1982 to work at another environmentally toxic site, the Hanford Nuclear Reservation in Washington State.[46]

Picking a Future

The sudden annihilation of its economic raison d'être presented those who remained in Kellogg with an opportunity few ever get to undertake—choosing the economic future of their community. Kellogg residents held a brainstorming session on the subject months before the last Bunker Hill employees left their posts. Suggestions included bottling black granular slag for sale and turning Kellogg into a gambling town. Many of the ideas, however, revolved around transforming the community into a tourist theme town. Suggested options included "western," "Alpine," and "mining camp."[47] No consensus vision emerged at the time, however.

In 1985, groups of Kellogg civic leaders traveled to Leavenworth, Washington, and Park City, Utah, to learn how those communities had successfully transformed themselves into destination tourist towns following the collapse of their traditional economic bases. Leavenworth, a onetime timber community that nearly died after the mill closed, revived its fortunes in the 1960s through its full-scale transformation into a mock Bavarian village. Park City, once a booming silver mining town, by the 1950s had fallen on hard times. The small community reemerged in the 1960s as a ski town that eventually boasted a gondola and easily accessible ski resorts. The civic leaders returned from their visits to Leavenworth and Park City full of inspiration. Following his viewing of Park City's gondola system, Kellogg Chamber of Commerce member John Reynolds enthused, "We learned that it can be done. Somebody else has done what we want to do and more."[48]

Unlike Park City, Kellogg did not need to build a ski resort. It already had one. The Jackass Ski Bowl had opened in the mountains above Kellogg and Wardner in early 1968, operating on 1,524 acres of land leased at no cost by Bunker Hill to the Fabulous Valley Corporation, an entity comprising primarily local investors.[49] Following Fabulous Valley's default on its Small Business Administration loans in 1973 and the subsequent foreclosure, Gulf-Bunker purchased the ski resort and renamed it Silverhorn. The company ran the ski hill through Bunker's demise, and the City of Kellogg took it over at the beginning of the 1983–1984 season.[50]

In 1985 Silverhorn represented Kellogg's greatest asset. But there was a problem. The seven-mile dirt road from Kellogg to the ski resort—whose base lay 3,400 vertical feet above town—was windy, treacherous, and slow. Ski resort consultants explained that there was no way to turn Silverhorn into a destination ski resort if the road remained the sole means of access to the ski hill. It was at this point that the gondola idea began to take shape, as civic leaders began discussing the concept of a long gondola to transport skiers from Kellogg directly to the base of Silverhorn. A feasibility study completed in the fall of 1985 concluded that the ski hill, coupled with the proposed gondola, "could generate 900–1,000 new permanent jobs."[51] Impressed by the way Park City and Leavenworth had used themes to successfully transform themselves, Kellogg leaders decided to marry the two examples and turn their community into a Bavarian-styled ski town. The idea was that a Bavarian theme would more effectively complement Kellogg's new identity as a destination ski resort than would the town's somewhat gritty, eclectic, mining town décor.

In December 1985, the Kellogg City Council passed an ordinance requiring that "Bavarian"-style lettering be used on all new signage in the town's business district. Additionally, all new construction would have to be done in "Alpine" style—with exposed beams, flower boxes, and murals painted on blank walls.[52] The following summer Kellogg voters approved offering $8 million in revenue bonds to finance gondola construction. Community support for the "if you build it, they will come" gondola project was nearly unanimous. The measure passed by a 97-to-3-percent margin.[53] Despite approval of the revenue bonds, securing funding for the gondola remained a long shot, in part because the bonds were to be repaid by receipts from the as-yet-nonexistent gondola. Kellogg sought assistance from the federal government for the $12 million project, pitching it to the Economic Development Administration and the Farm Service Agency, and seeking an urban development action grant and a community development grant.[54] The city struck out in all cases. In the end it was a far older form of federal funding, pork-barrel spending, that came to the rescue. Senator McClure managed to insert a $6.4 million grant for the gondola into the $600 billion omnibus appropriations bill that Congress approved in late 1987, tucking it into the appropriations for a US Forest Service program to pay for the development and management of forests.[55] The federal grant, however, had to be fully matched by September 30, 1988, or it would disappear. Forced once again to scramble for money, Kellogg secured matching funds from VonRoll Transport Systems, the Swiss manufacturer that won the bid to construct the gondola, within weeks of the federal deadline.[56] In order to cement the deal with VonRoll, however, the citizens of Kellogg agreed to back their revenue bonds with an additional special tax levy of $2 million over the next twenty years.[57] Kelloggers had mortgaged their future for a gondola.

Ski Town

June 30, 1990, was a day of celebration for Kellogg. That was the date for the official opening of the gondola and for Silver Mountain, the renamed and enhanced Silver-

FIGURE 7.1. The view from Gondola Village, looking up toward Silver Mountain (photo by the author, March 2010).

horn.[58] Appropriately enough, Senator McClure was on hand as a special guest for the grand opening of "America's newest ski and summer resort."[59] The centerpiece of the resort was the 3.1-mile-long gondola, the longest in North America. It ferried passengers up 3,400 vertical feet from the emerging Alpine-styled base village on the western end of Kellogg (not far from the old smelter site) to the Mountain Haus upper terminal, in the shadow of Kellogg Peak. In a fitting symbol of the town's changed economy, the gondola would carry skiers and sightseers along the same path where in the 1880s and 1890s Bunker Hill's tramway hauled ore from the mine to the Kellogg mill. On its 500 acres of skiable terrain, Silver Mountain offered 2,200 vertical feet of skiing divided into 38 runs. In addition to construction of the gondola, three connecting lifts had been added to the resort.[60] Although Silver Mountain had plans to expand, its size placed it in the middle range of US ski areas, a far cry from the size and scope of such famous destination ski areas as Telluride, Vail, and Park City.[61] Owned jointly by the City of Kellogg, VonRoll, and the Hagadone Hospitality Corporation, Silver Mountain would be managed by Coeur d'Alene–based Hagadone Hospitality. Hagadone hired Peter Forsch, a manager for seventeen years at the renowned Aspen Ski Resort, to oversee the resort's development and operations.[62]

Much of the early buzz surrounding the new resort was enthusiastic. Envisioning a year-round recreation hub with summer hikers, mountain bikers, and mine tourists replacing winter snowboarders and skiers, Silver Mountain director of sales Todd Lund forecasted, "We're going to be just like Knott's Berry Farm, where you have tons of activities going on."[63] While admitting that Kellogg lacked the accommodations to house most of the visitors that would be generated by Silver Mountain, Forsch predicted that the resort's presence would spawn a real estate boom throughout the Silver Valley.[64] Encouraged by the gondola's strong first summer use, Silver Mountain's owners announced plans to inject an additional $5.3 million into the resort over the next six months. The capital would be put toward building a three-thousand-seat amphitheater near the gondola's upper terminal, adding forty additional gondola cabins, purchasing snow-making equipment, and installing a lighting system for night skiing.[65]

Wilkommen zu Kellogg

While the new ski resort received mostly positive press, the same was not always the case for the "new Kellogg," replete with its Bavarian stylings. Some mocked the town's new welcoming sign, which read "Wilkommen zu Kellogg." In an Associated Press article, *New York Times* columnist Timothy Egan described Kellogg as "slouching toward trendiness," before concluding that the town was "trying to avoid a future as a ghost town by embracing the quaintness of a Bavarian village."[66] More harshly, a *Spokesman-Review* reporter wrote, "When local boosters chose a Bavarian theme to turn this depressed mining town into a fairy-tale tourist village, they had in mind a Cinderella transformation. . . . After four years, what they've got is more like Beauty and the Beast."[67] Noting that fewer than a quarter of downtown Kellogg's businesses had "gone Bavarian" and that "even these aren't entirely convincing," the reporter's assessment was that "the 'theme town' concept . . . faces a serious challenge in this northern Idaho town . . . Kellogg still resembles more the rough mining town it is than the charming village it wants to be."[68]

In truth, not everyone in Kellogg was fully on board with the "Bavarian" idea. Some insisted that the town should proudly reflect its mining and smelting heritage rather than adopting a false front. Katie Lutz, a merchant who had worked in the Bunker Hill Mine, expressed such concerns, commenting, "I have some real reservations about going Bavarian. . . . We should never forget that we're mining people in a valley built on the mines."[69] In the early 1990s, city efforts to punish Kellogg merchants who were not complying with a sign ordinance that mandated the use of Bavarian lettering in all commercial signs touched off controversy over both the restrictive nature of the ordinance and the Bavarian theme itself. The sign ordinance, touching on the deeper issue of community identity, divided Kellogg. Dale Brown, a local gift shop owner who supported the ordinance, argued that the Bavarian theme represented Kellogg's best opportunity for economic development and maintained that the sign ordinance was supported by "the majority of business owners" in town.[70] On the other hand, Jack De Feo, owner of an Italian diner, believed the town should promote its western and mining heritage and felt the Bavarian theme was "a slap in the face to the miners in this valley."[71] Other merchants complained that the ordinance was overly restrictive and too expensive to comply with. Dave Smith, owner of Dave Smith Motors, a car dealership that was one of the Silver Valley's largest employers, branded the Bavarian lettering illegible and promised to fight it "with every legal means at our disposal."[72]

On Silver Mountain

Silver Mountain's inaugural ski season started strong, perhaps spurred by curiosity about the gondola, but eventually faded. Based upon early season activity and overcrowding at the area's facilities, Forsch predicted the hill would record one hundred thousand skier visits for the 1990–1991 season. The resort's failure to reach that target was met with disappointment by some Kelloggers.[73] Despite the investment of another million dollars following the initial season, over the next few years Silver Mountain failed to reach the level of the top national destination ski resorts, or even the top tier of regional ski hills. While, for example, Big Mountain, a leading regional resort, recorded 295,000 skier visits for the 1995–1996 ski season, Silver Mountain managed only 95,000. This represented little or no improvement over the resort's initial season.[74] While its level of use placed Silver Mountain in roughly the same category as nearby ski areas Schweitzer Mountain and Mount Spokane, the Kellogg resort remained worlds removed from a Vail or an Aspen, both of which averaged over a million annual skier visits by the early 1990s.[75] In May 1996, following yet another lackluster ski season, Silver Mountain was sold to Eagle Crest Partners, a resort-management subsidiary of Oregon's Jeld-Wen, a major manufacturer of windows and doors.[76] The news worried Kelloggers, in part because the terms of sale permitted Jeld-Wen to sell or move the gondola. To date, however, this has not come to pass—the gondola remains a fixture of the middling ski resort.

Resort Community

A destination ski resort is more than simply lifts, runs, and skiers. A big part of the experience of such a place for visitors, and a significant source of revenue for resort owners, comes in the form of other amenities. Eateries and bars, clothing stores, art galleries, lodges, golf courses, and real estate opportunities ranging from tiny condominiums to palatial homes have become part and parcel of the contemporary ski resort scene. While it took a number of years for Silver Mountain to engage in this side of the business in earnest, by the mid-2000s it had jumped in with both feet, joining the real estate frenzy of the era. In 2004, Jeld-Wen's construction and real estate arm, Jeld-Wen Communities, broke ground on phase one of its Morning Star Lodge condominiums at Gondola Village (the renamed Base Village), and quickly sold out the sixty-eight units.[77] The following year the phase-one condominiums were opened for use, phase two of the Morning Star project was announced and 104 units were sold, and ground was broken for Silver Rapids, an indoor waterpark in Gondola Village. Development continued apace in 2006, when phase three's ninety-nine condominiums sold out in little time.[78] In 2007 construction began on Galena Ridge, an eighteen-hole golf course and real estate development west of Gondola Village on the site of the former Bunker Hill works.

Kellogg seemed finally to be emerging as the recreation community it had been striving to become. Tourists and investors began buying up older homes for ski rentals and vacation cabins.[79] Long-depressed real estate prices in town climbed. Although it was hard for Kelloggers to fathom the idea that local condominiums were selling for as much as $800,000, by the mid-2000s the town was emerging as a draw for winter recreationists seeking a second home that was significantly more affordable than what could be found in places such as Park City or Lake Tahoe, California.[80] Condo-flipping commenced. Developers talked of building as many as three thousand new condominiums and single-family dwellings.[81]

As it happened, the timing of Kellogg's real estate development boom hardly could have been worse. In the summer of 2007 the national housing bubble burst and the fall of 2008 witnessed the beginning of the worst economic slowdown to grip the United States since the Great Depression. Not surprisingly, the pace of construction in Kellogg slowed dramatically. By April 2008, chain link fences surrounded two stalled condominium projects and Silver Mountain had dropped its asking price for new townhouses from $585,000 to $395,000.[82] Nine holes of the proposed eighteen-hole Galena Ridge golf course finally opened in 2010, years behind schedule. It remains unclear whether the remaining nine holes will be built.[83] Fortunately for residents of Kellogg and other parts of the Silver Valley, a simultaneous and unexpected resurgence in the price of silver led to renewed activity in the "old West" sector of the area's economy that somewhat buoyed it.[84] With the price of silver rising above seventeen dollars an ounce, the valley's three remaining mines quickly added two hundred miners to their payrolls. Paying an average of $57,000 a year, the County's seven hundred mining jobs sustained a far higher standard of living than did those of the assorted ski lift operators, maids, waiters, baristas, and others employed in the area's tourist economy.[85]

A Superfund World

In 1983, a seven-by-three-mile stretch of the Silver Valley, centered on the former Bunker Hill smelter, was designated a Superfund site and placed on the EPA's National Priorities List. The twenty-one-square-mile area, known as "the Box," encompassed the communities of Pinehurst, Page, Smelterville, Kellogg, and Wardner, and was home to about seven thousand people.[86] Although by 1983 Silver Valley residents had already experienced more than a decade's worth of contact with the EPA, the relationship between the north Idahoans and the federal agency was about to enter a new and far more intensive stage.

The first phase of the cleanup was dedicated to minimizing the health threats posed to children by soil lead in and around the site's communities. Amid wrangling over who would pay for the cleanup and apparent foot dragging by Robie Russell, the EPA administrator who ran the agency's Pacific Northwest Division from 1986 to 1989, the yard cleanups that lay at the heart of this effort did not commence until 1989.[87] By 2008, when the first phase was completed, over 2,700 residential yards had been excavated and had their metals-laden soil replaced with clean dirt. In a sign of success, area children's blood lead levels dropped with the soil removal, and now sit at approximately the national average.[88]

The second phase of the EPA cleanup centered on the old Bunker Hill works. This Herculean task involved the hauling of about 2.4 million cubic yards of mine tailings and contaminated materials, and the demolition of more than two hundred buildings in the industrial complex. The waste was buried in the 60-foot-high, 260-acre central impoundment area just west of Kellogg that runs for over a mile along the south side of Interstate 90. The site's mercury was encased in concrete and the rest was wrapped in a thick plastic sleeve before being capped and covered with dirt.[89] In 1995 Kellogg lost two of its most important industrial landmarks when the EPA dismantled and razed the lead smelter and zinc refinery. The following year some locals lost their battle to preserve as tourist attractions and historical monuments the tall stacks built twenty years earlier to vent smelter and refinery gases.[90] On Memorial Day, 1996, locals and tourists turned out to watch as the towers were blown up and toppled.

EPA: Friend or Foe?

In the years since the original Superfund designation, the EPA has been a lightning rod in the Silver Valley. The storm surrounding the agency intensified, however, following its decision in 1996 to expand the coverage area from 21 to roughly 1,500 square miles focused on the Coeur d'Alene River basin. In that year, the EPA also joined a US Department of Justice suit against Silver Valley mining companies for natural resource damages in the Coeur d'Alene basin and reneged on a pledge not to expand the Superfund project beyond the Bunker Hill "Box."[91] In September 2002 the agency formally announced its plans to expand the cleanup project to cover the one-hundred-plus miles between Mullan, Idaho, and Spokane, Washington. The basin cleanup plans called for the expenditure of an estimated $357 million over the next thirty years.[92] The EPA's reversal tapped a vein of anger in the Silver Valley, where many viewed the Superfund designation as a stigma that hampered efforts to build a tourist economy and the agency's presence as a barrier to the continued practice of mining.[93] In the early 2000s, when anti-EPA signs dotted the area, nearly three in five surveyed Silver Valley residents rated the agency poorly for gaining their trust.[94] The EPA itself described the Silver Valley cleanup as "unusually contentious."[95] Following the admonition of a *Shoshone News-Press* columnist that locals should be required to arm themselves against federal agents seeking access to their property, uniformed police and undercover officers accompanied EPA officials to their next Silver Valley meeting.[96]

Attacking the "Local Hero"

Despite the opposition of many valley residents to the expansion of the Superfund site, some locals supported the agency's decision. As Barbara Miller, the leading voice of the Silver Valley People's Action Coalition, an organization formed in 1986 to advocate for children's health and a full cleanup of the valley, said succinctly, "The EPA is our only hope for this community being cleaned up."[97] Miller and her allies contended that serious health problems remained in the Silver Valley and that the only realistic opportunity for remedy lay in the further scouring of decades worth of mining and smelting wastes.

The controversy surrounding Miller's nomination in 2001 for a Ford Foundation award emblematizes the tensions in the valley over the federal cleanup. Nominated by a Silver Valley woman who remained anonymous fearing reprisals, the Kellogg activist became one of thirty-six finalists for a Ford Foundation honor that recognized people worldwide who had improved life for others.[98] In response to the nomination, *Shoshone News-Press* editor Dan Drewry wrote an editorial encouraging locals to send the Ford Foundation letters opposing Miller's receipt of the award. Drewry maintained that "we, as a community, know Barbara Miller and her claque represent a tiny, disaffected minority. You have a chance today to help choke off the stream of lies, half-truths and distortions that emanate from the organization [the Silver Valley People's Action Coalition]." In a community reaction that a Ford Foundation official described as "highly unusual," the foundation received a flurry of letters of opposition to Miller's nomination.[99] Nonetheless, the foundation named Miller one of its twenty award winners and granted her $130,000 to continue her efforts.[100] Kellogg mayor Roger Mangum's response to Miller's recognition was hardly enthusiastic. "I don't think we'll have a parade," said Mangum. "It's a slap in the face. Talk to 1,000 normal people and they'd be surprised to see she was even considered for it."[101] In an implicit acknowledgment of the threat of retribution against Miller, however, Mangum urged townspeople to remain peaceable in their treatment of the activist, saying, "I'll just remind people, especially after last week [the events of September 11, 2001], that we have freedoms in America. . . . Don't tread on others. Let's not do anything."[102]

EPA, Economic Engine

In 2009, as part of a nationwide settlement with the US Department of Justice, international mining and smelting giant ASARCO, Inc. (legal successor to the Federal Mining & Smelting Co., which owned valuable area properties such as the Tiger-Poorman, Morning and Standard-Mammoth Mines, and the Page Mine and Mill), handed over $494 million for EPA cleanup efforts in the Coeur d'Alene basin.[103] In June 2011, Coeur d'Alene–based Hecla Mining Company, owner of the Silver Valley's Lucky Friday Mine, announced the settlement of its federal lawsuit for $263 million. Approximately $197 million of that would go toward the EPA's continuing basin cleanup.[104] Taken together, the settlements represented an infusion of nearly $700 million into the EPA's coffers. In 2010, the agency began taking public comment on another proposed expansion of the Superfund project, projected to cost $1.3 billion and take between fifty and ninety years to complete.[105] The stated aims of this next, most lengthy, proposed phase of the cleanup were: preventing repollution of the basin via flooding, protecting groundwater sources, cleaning up abandoned mine sites, continued progress vis-à-vis human health, and enhanced protection of wildlife.[106] Silver Valley skeptics postulated that the huge ASARCO settlement, more than sound scientific rationale, drove the EPA's decision to push for another expansion and extension of the cleanup project.[107]

Although anti-EPA signs no longer dot the valley, the agency's decision to continue its cleanup efforts for decades more has met with widespread local condemnation. Local and state officials appear to be united in their opposition as well. When in the fall of 2010 the EPA held hearings in Wallace for the Record of Decision (ROD), the official blueprint that will guide its work on the next phase of the Superfund cleanup, seven Silver Valley mayors, area state legislators, and a representative of Governor Butch Otter all showed up to express their disaffection with the plan.[108] Numerous speakers expressed the view that the "listening sessions" were a waste of time because the federal agency did not really consider locals' opinions in making their decisions.[109] In an editorial written shortly before the hearing, local state representative Dick Harwood said, "Public opposition to this plan is nearly unanimous . . . their plan is too expensive, too big and the wrong approach."[110] With characteristic floridity, Drewry suggested that the EPA was "a crazed bureaucracy throwing money at a non-existent problem," and urged, "Our elected [congressional] representatives, by reining in [the] EPA, could save hundreds of millions of tax dollars. Work a little Tea Party magic on EPA, gentlemen."[111] Though unlikely to mollify its more ardent valley critics, the EPA announced in early 2011 that, in response to the nearly six thousand local comments the agency had received for its upcoming ROD, it was giving serious consideration to trimming $300 million from the project's proposed $1.3 billion budget.[112]

While the wisdom and efficacy of aspects of the EPA's thirty-year cleanup effort can be argued, it is hard to gainsay that the agency has spent a good deal of money in the economically depressed Silver Valley during its tenure there. The unpopular federal agency, in fact, almost certainly has been the leading economic driver in Kellogg, and probably the entire valley, over the past twenty-five years, assuming the role once played by "Uncle Bunker." As of 2002, Superfund spending accounted for $77 million in jobs and community improvements in Kellogg and other Silver Valley towns, with $42 million of that used to hire locals.[113] Some six hundred contractors worked the cleanup during that period, and 85 to 90 percent of the labor force was local.[114]

With Gulf Resources' bankruptcy in 1993, the EPA assumed financial responsibility for the Superfund cleanup. Under the guidelines of the Davis-Bacon Act, the agency and its subcontractors on the project were required to pay prevailing union-scale wages. About five hundred Silver Valley workers thus made between twenty-two and thirty dollars an hour on the cleanup—roughly two to three times what they otherwise would have earned—while also receiving union medical and pension benefits.[115] While adherence to Davis-Bacon undoubtedly inflated the cost of the cleanup, it also meant significant money and benefits for Silver Valley workers. In addition to its direct injection of cash and jobs into the valley's economy, the EPA also provided the City of Kellogg with ten thousand dollars' worth of computers for its schools and libraries, a $230,000 "greenbelt" that transformed the area adjoining the former rail line through town (now a bike path) into a park-like space, $542,000 in street maintenance funds, and $17 million in flood control and street projects.[116] Though some argued that Superfund jobs were not particu-

FIGURE 7.2. The Trail of the Coeur d'Alenes at Mullan, Idaho (photo by the author).

larly helpful economically because they tended to be seasonal—and certainly they could not compare with the area's lost mining and smelting jobs—in the Silver Valley's postindustrial era the jobs represented some of the best opportunities to be found.[117]

Bike Paths to the Future

A telling example of the Silver Valley's transition from an industrial to a recreational economy is the transformation of two of the area's former railway beds into bike paths. In 1998, the Route of the Hiawatha Trail opened on the Montana-Idaho border near Lookout Pass. The bike trail ran along the old right-of-way of the Milwaukee Railroad, a perpetually troubled line that went out of business in 1980. In 2001, the trail added the Taft Tunnel, a 1.66-mile tunnel through the Bitterroot Mountains on the Idaho-Montana border. The fifteen-mile-long Route of the Hiawatha Trail, considered by some to be the "crown jewel" of the national rails-to-trails system, includes nine tunnels, seven high trestles, and abundant mountain scenery.[118] Lookout Mountain Ski Area manages the trail and does a brisk seasonal business renting bike equipment and running the shuttle bus service that assists the many who want to ride the trail as only a one-way, mostly downhill trip. The Route of the Hiawatha has proven to be extremely popular, especially since the opening of the Taft Tunnel, drawing many recreation-seeking families and individuals to the Silver Valley. In 2010, for example, the trail attracted 11,321 riders in July and another 12,263 in August, shattering usage records set the previous year.[119] The trail likely will only become more popular when an additional thirty-three-mile stretch extending it to Saint Regis, Montana, is completed.

The Route of the Hiawatha Trail is not the area's only popular bike path, however. Since 2004, its first official year of use, the seventy-three-mile-long Trail of the Coeur d'Alenes has been ridden by about a hundred thousand cyclists annually.[120] The trail runs from Mullan, in the mountains on the east end of the Silver Valley, to Plummer, in the hilly Coeur d'Alene Indian reservation lands west of Lake Coeur d'Alene. Those who manage the long trek will ride through and around mountains, defunct and active mines, the old Bunker Hill works, the lush wetlands of the lower Coeur d'Alene River, and picturesque Lake Coeur d'Alene. The ribbon of asphalt now carries cyclists over the path where heavily laden Union Pacific trains once ferried lead, silver, and zinc to distant markets. Over the years those trains spilled large quantities of toxic metals, so camping along the trail is strictly prohibited. Numerous trail signs remind riders to stay on the asphalt.[121]

In April 1993, Union Pacific stopped running trains to the Silver Valley. To preserve its 150-foot right-of-way, the railroad proposed building a recreational trail on its former line under the guidelines of the federal Rails-to-Trails Act. Under an agreement with the EPA, the State of Idaho, and the Coeur d'Alene Indian tribe, Union Pacific removed its track and other features, scraped away lots of contaminated dirt, built protective barriers, and capped the railbed with a 2.5-inch layer of asphalt. The railroad also set up a $2.6 million escrow account for trail maintenance.[122] Although the trail is far from pristine—hence the frequent warning signs—the EPA has deemed it safe for cyclists who follow the rules.

Perhaps surprisingly, this bike path through a Superfund site has proven highly popular and has become a significant economic engine for the Silver Valley. Visiting cyclists who ride the trail tend to be from higher income brackets, and they spend money on local gas, food, and lodging during their excursions. Jon Ruggles, president of the group Friends of the Coeur d'Alene Trails, estimated the trail's local economic impact at $6 million during the 2004 season.[123] Discussing the Trail of the Coeur d'Alenes, Mike Domy, owner of the Excelsior Bike Shop in Kellogg, said, "We're becoming a playground for people in Seattle and Portland. . . . [The trail] helps us round out our tourism economy. It's not just skiing anymore."[124]

Over the past two decades the bike trails and Superfund spending have emerged as the only reliable bright spots on Kellogg and the Silver Valley's economic horizon. Mining remains dangerously susceptible to price fluctuations in

FIGURE 7.3. A warning sign posted on the Trail of the Coeur d'Alenes (photo by the author).

metals. Skiing and snowboarding, while important for winter economy, have not proven to be the recreational draw their supporters had hoped for. Silver Mountain has not become a major destination ski resort and is not well known outside the Inland Empire. The real estate and construction boom of the mid-2000s that centered on Gondola Village came to a screeching halt with the bursting of the national housing bubble. It is unclear whether the resort-based development on the west end of Kellogg will eventually revive.

Both Shoshone County and Kellogg continued to lose population in the first decade of the twenty-first century. The county's total dropped by 7 percent to 12,765, and its largest city shed 11.5 percent of its population to end the decade with 2,120 residents, a decline of nearly two-thirds since the late 1950s.[125]

Although the Kellogg real estate boom helped boost the county's median household income by 31 percent between 2003 and 2008, the subsequent bust led to a 9 percent drop in 2008–2009.[126] In August 2011 nearly one in six job-seeking adults in Shoshone County was without work.[127] The local mining industry, resurgent in 2007–2008, witnessed reversals triggered by the global recession in late 2008 and early 2009 that led to the closure of the Sunshine Mine and the loss of over two hundred mining jobs. In 2010, however, the resurgent price of silver led to the employment of additional miners at the valley's two operating silver mines, the Galena and the Lucky Friday. In that year Shoshone County's 539 miners earned an average of $70,919—well over twice the county's average wage.[128] The death of veteran Silver Valley miner Larry "Pete" Marek from a cave-in at the Lucky Friday in April 2011, however, served as a reminder of mining's dangers.[129]

Thirty years after the fact, Kellogg's (and to an extent Shoshone County's) decision to pursue recreational tourism as its principal economic strategy has produced rather thin gruel. In 2010, the 407 Silver Valley residents who worked in the leisure and hospitality industry represented less than 8 percent of the county's total workforce, only marginally more than the nearly 350 employed by Kellogg's Dave Smith Motors.[130] Those who work in recreation, unfortunately, do not earn much for their efforts. In 2010, when the county's per capita income stood at $30,811, the average wage for one of the area's leisure and hospitality jobs was $12,106, or below the federal poverty line for a married couple.[131] Shoshone County's postindustrial, tourist economy jobs paid slightly more than one-sixth of what its remnant industrial economy, mining jobs, did. It is no wonder that many locals yearned for a return of large-scale mining to the Valley. While advanced modernity, as theorized by Beck, can support both green values and a vibrant, diversified economy in large cities such as Seattle or Boston, or in successful tourist communities like Aspen or Park City, in former single-industry towns such as Kellogg that struggle to find a sustainable economic basis, nostalgia for the bygone classical industrial stage, despite the environmental and social pitfalls that form of political economy entailed, is understandable.

In spite of the profits generated, little of the proceeds of Kellogg's mid-2000s real estate boom appear to have trickled down to uptown Kellogg. Though Uptown's commercial district had been the throbbing heart of the small city during the Bunker Hill era, Kellogg's switch to a tourist economy shifted the town's geographic center to Silver Mountain's Gondola Village, approximately half a mile northwest of Uptown. The 2008–2011 recession hit all of Kellogg hard but its results were particularly noticeable in Uptown, where boarded-up windows and abandoned buildings, along with vacant storefront windows displaying "For Sale"

FIGURE 7.4. Kellogg YMCA (photo by the author, March 2010).

and "For Rent" signs, became familiar landscape features.[132] In September 2010, Kellogg business leaders held a town hall meeting at the former Steelworkers building to generate ideas for reviving the town. In language reminiscent of the efforts of the early 1980s to develop a vision for the future of the town, Brenda Stinson, an Uptown merchant spearheading the effort, encouraged creative participation in the forum, saying that "any ideas will be welcome, no matter how crazy."[133]

At the town hall meeting some expressed concern over the blight that was overtaking Uptown. A number of attendees voiced the view that the city should embrace its identity as a mining and smelting community and discard the long-languishing efforts to become a faux Bavarian village. One resident, Michael Murphy, exhorted simply, if not with geographic exactness, "We're not Austria . . . this is not Austria."[134] Amplifying upon the theme, Stinson suggested that Kellogg should draw upon its heritage as "the only town in the U.S. founded by a jackass," and host a variety of tourist events based upon the jackass theme, such as an annual Jackass Days, a "running of the jackass," and the naming of a local "jackass of the year."[135] Although a series of town hall meetings was held in 2010, a new vision for the direction of the community did not emerge.[136]

As the current state of Kellogg and continuing struggles of Kelloggers to craft a viable future for their town attest, the transition from an industrial to a postindustrial community is fraught with challenges. More than thirty years into the process, Kellogg and the Silver Valley still wrestle with fundamental questions of identity, direction, and viability. Economic restructuring has proven far more difficult than all the initial hype and hope surrounding Bavarian signs, resort development, or a gondola indicated. As it happens, if you build it, they might or might not come. And even if they do come, they likely will not support pay scales comparable to those of the community's former basic industry. Although Kellogg and the rest of the Silver Valley undoubtedly are far cleaner and healthier places to live than they were thirty years ago—and this is no small thing—the communities are hardly as strong or vibrant as they were when mining and smelting ruled the valley. Difficult irony abounds. The malignant smoke is now gone, but with it too has departed much of the life and livelihood of Kellogg and the valley.

CONCLUSION

The legacies of the Coeur d'Alenes are outsized. A place that has produced more silver than any other in world history, it also has been one of the largest producers of lead and zinc in US history—all told, more than twenty-six billion dollars' worth of metals.[1] Home to the nation's three richest silver mines—the Sunshine, Galena, and Bunker Hill—the area also boasted sizable lead smelting and silver and zinc refining facilities. For much of the twentieth century, the Coeur d'Alenes' mines, mills, smelter, and refinery employed thousands of workers and supported a vibrant working-class culture. But the Coeur d'Alenes also have an alternate, albeit intimately related, legacy of environmental degradation and despoliation that has resulted in its status as one of the nation's largest and most challenging Superfund sites over the past thirty years.[2] Over fifty-six million tons of mine tailings were sent down the South Fork of the Coeur d'Alene River by mining and milling companies, polluting an estimated twenty-five thousand acres of farmland with lead and other heavy metals by 1900 and rendering the Coeur d'Alene River and its tributaries nearly devoid of aquatic life by the early 1930s.[3] Until the late 1960s, the mines and mills of the Coeur d'Alenes essentially used the South Fork of the Coeur d'Alene River as an industrial sewer to transport wastes away from their worksites. Even today, after more than forty-five years of tailings impoundment that meets the industry standard and thirty years of cleanup efforts, roughly twenty miles of the streams cannot sustain a reproducing fish population, and about ten miles of tributaries have virtually no aquatic life. In addition, lead poisoning is responsible for a large number of waterfowl deaths each year in the lower Coeur d'Alene River basin: more than fifteen thousand acres of wildlife habitat contain sediments and soils that are acutely toxic to waterfowl.[4]

A perhaps even more destructive environmental legacy, however, can be traced to the Coeur d'Alenes' history of lead smelting and zinc refining. Visitors to the area in the middle decades of the twentieth century routinely noted the denuded hillsides surrounding Kellogg. The lack of vegetation owed principally to the large quantities of sulfur dioxide emitted by the Bunker Hill Company's zinc refinery and lead smelter. The acrid, foul-tasting gas killed trees and other plants, and on days of particularly strong concentrations, impaired human health. The smelter's excretions of lead fumes, however, represented an even more serious threat to human health. Even before the 1973–1974 lead health epidemic, many Kellogg area residents undoubtedly suffered from chronic lead poisoning due to their exposure to the lead smelter's noxious vapors. In the wake of the massive emissions that followed the baghouse fire of September 1973 and Gulf-Bunker's ill-fated decision to continue running the smelter at full capacity with a badly damaged emissions-filtering system, acute lead poisoning became the rule for those living within 2.5 miles of the smelter stack. This was particularly damaging to the health of young children, who as a group face significantly heightened health risks from lead intoxication. There is significant evidence to indicate that a good many children who lived near the smelter in 1973–1974 suffered real and permanent harm from the experience. Tests conducted in the mid-1970s showed evidence of slowed nerve conduction velocity and anemia in this group. A mid-1990s follow-up study revealed a host of continuing maladies affecting a number of the former "baghouse kids"—these included anxiety, infertility, learning disabilities, high blood pressure, anemia, diminished hand-eye coordination and grip strength, slowed nerve conduction velocity, poor vocabulary, depression, memory loss, and trouble concentrating.[5]

Another group affected by the pollution of the air in the Coeur d'Alenes consists of lead plant workers, and, to a somewhat lesser degree, zinc plant workers, both of whom consistently bore the burden of the community's highest levels of exposure to aerosolized lead and other toxic heavy metals. Although the problem of occupational lead poisoning was already well known by the time of the construction of the Bunker Hill smelter in 1917, the "leading" of workers remained a significant problem at the smelter throughout its history. Health and mortality studies of Bunker Hill lead smelter workers conducted from the 1970s through the 1990s demonstrate that as a group they paid a steep price for working at the facility. In addition to peripheral nerve dysfunction, lengthened choice reaction time, increased feelings of hostility and depression, and disordered calcium metabolism, the studies charted elevated rates of renal disease, renal failure, and increased mortality from these causes and from stroke in the Bunker Hill smelter workers.[6] With their bodies at the "biological core" of the smelter's workscape, the workers' elevated rates of renal disease and other severe health issues offered evidence of a dangerously toxic workplace.[7] Smelter workers' sickened kidneys bespoke a heavily leaded work environment.

The mining, milling, and smelting of the Coeur d'Alenes has enriched numerous corporate managers and shareholders and sustained thousands of workers living in the string of working-class communities that runs from Mullan to Kingston, Idaho. The industry long provided substantial economic support to more distant centers such as Coeur d'Alene and Spokane. Since 1885, the area's prodigious production of base metals such as lead and zinc and a precious metal, silver, also has played a role in the development of global industrial civilization. Lead from the Coeur d'Alenes has found its way into the world's supply of a great many products, including electrical cables, X-ray shielding vests, paint, gasoline, and batteries. The area's zinc has been integral to the production of die castings, alloys such as brass and aluminum solder, and zinc oxide. And Coeur d'Alenes' silver can be found in electrical conductors, jewelry, coins, and photographic film the world over. The social costs of this production, however, have been significant, to say the least. These run from the incalculable losses of potential and opportunity for the children affected by the 1973–1974 lead health epidemic to the more quantifiable deaths of fish, geese, swans, horses, cows, trees, and bushes that wound up in the path of the mining district's tailings, smoke, and fumes. They range from the leaded paint and gasoline that coated the planet during the nineteenth and twentieth centuries—decreasing the intelligence of millions and killing thousands—to the renal disease and early death suffered by Bunker Hill smelter workers, and from the ongoing health issues faced by those who grew up in the shadow of the smelter to the millions of tons of toxic heavy metals that lie like an inverted Sword of Damocles at the bottom of Lake Coeur d'Alene, threatening its ecological future.

For the most part, these costs have been shifted. As has been the rule under capitalism, instead of being shouldered by the companies responsible for creating them, or their consumers, the social costs of industry have been borne largely by the local environment and by the bodies of relatively powerless groups of people—in the Coeur d'Alenes this principally has meant Bunker Hill smelter workers and locals living near the smelter. When it comes time to pay for the cleanup of a polluted environment, or the medical treatment of a sickened worker, the company responsible may not be around to be held liable, as has been the Silver Valley's experience with Gulf Resources, which declared bankruptcy in 1993.[8] In a corollary to what has been termed "the fisherman's problem," private individuals or corporations have little incentive to refrain from polluting the public commons (e.g., air and water)—the gains from doing so will accrue to them individually, while the costs will be borne by society generally.[9] The history of human health and the environment in the Coeur d'Alenes fits this rule. Although Bunker Hill and the other large companies in the district made some, relatively minor, efforts to protect the environment and worker health before the "environmental turn" of the 1960s–1970s, the bulk of their efforts came after federal and state law, backed by regulatory agencies, compelled them to do so.

It almost goes without saying that not all industries are created equal. Some industries, for instance, tend to offer relatively high wages, while others tend to be low-paying. And some industries tend to be more healthful for workers than others. By almost any measure, lead smelting is a dirty and unhealthful industry. There were, however, significant differences between lead smelters even within this "dirty industry." Owing to different corporate attitudes and priorities, some smelters were relatively healthier places to work than others. By the 1970s, if not earlier, the Bunker Hill smelter had fallen significantly behind the US industry standard for cleanliness. Health experts who visited the facility during this period routinely commented upon its lack of adequate hooding and venting equipment, relative dustiness, and need for improved "housekeeping." Bunker Hill's long-standing policy of making smelter workers responsible for their own health—represented in the mid-1970s by the requirement that all smelter workers wear respirators at all times in the facility—permitted the company to save money that it might otherwise have spent on cleaning up the smelter environment. By the 1970s, Bunker Hill faced an uphill battle to bring air lead levels at its aging smelter into compliance with increasingly demanding federal requirements. Had the company kept pace with the industry standard for smelter cleanliness, as was done at Doe Run and other facilities in Missouri's historic "lead belt" and at the Ronnskar plant in Sweden, its smelter workers would have been healthier and the cost of compliance with new workplace rules far more manageable.[10]

In another missed opportunity, the major mining and milling companies of the district, as represented by the Mine Operators Association (MOA), in 1931 opted not to invest in the tailings disposal system recommended by fisheries biologist M. M. Ellis and supported by two of the three members of the Coeur d'Alene River and Lake Commission. This proposed tailings system was modeled upon one employed in Kimberley, British Columbia, by Sullivan Mines and Mills, a lead and silver mining and milling company. The Kimberley system used wooden flumes to transport mine wastes to a number of settling basins. There, mill tailings settled out to the extent that plankton, aquatic insects, and fish life could be observed at the point where the "tailings water" reentered the Saint Marys River.[11] Although Ellis warned that the MOA's preferred option, merely the employment of a suction dredge at Cataldo, would fall far short of restoring the ecological health of the Coeur d'Alene River and Lake Coeur d'Alene (and of course would do nothing for the upstream South Fork), this nonetheless was the course that the district's principal mine and mill owners followed. While it might have been more expensive than the dredge, installing a "Kimberley system" in the early 1930s would have significantly limited the ecological harm to the Coeur d'Alene basin's ecosystem and begun the restoration of plant and animal species more than three decades earlier than it occurred.[12]

Another example of the significance of corporate decisions for local environments involves the Consolidated Mining and Smelting Company (COMINCO), a major Canadian mining and smelting firm that operated a lead-zinc smelter at Trail, British Columbia. In the late 1930s COMINCO adopted a supplemental control system (SCS) that significantly reduced both sulfur dioxide emissions at its smelter and the resultant damage to vegetation.[13] The Canadian company was acting in response to a mandate from the International Joint Commission (IJC), a US-Canadian body established to resolve a dispute over smoke damage to the crops of Northport, Washington, farmers from the nearby Canadian smelter. Although by the late 1940s locals and US Forest Service officials were complaining vigorously about the Bunker Hill smelter and zinc plant's sulfur dioxide damage to area vegetation, the company did not adopt a COMINCO-style SCS until the 1970s.[14] Once again, although technology existed that would have enabled Bunker to minimize its damage to the local environment, the company chose cost savings over the implemention of cleaner technology.

The ultimate example of Bunker Hill's conscious cost shifting of its pollution to human health and the environment undoubtedly is the 1973–1974 lead epidemic, where the company made the Faustian decision to run its smelter at full bore for months even though much of its air filtration system was inoperative. From January to March 1974, the company compounded its wrongs by venting unfiltered smelter fumes and gases directly into the atmosphere, bypassing what remained of the baghouse and sending its toxic fumes directly up the smelter's main stack.

The history of the Coeur d'Alenes is part and parcel of the story of industrial modernity, and its stepchild, risk. Until the 1880s, the future mining, milling, and smelting center lay in the midst of a remote, mountainous area that would have been viewed by most Americans of the time as the rawest and most peripheral sort of wilderness. Quartz mining, with its nearly unmatched capacity to rapidly generate great wealth—and thus to quickly summon vast quantities of labor, technology, and capital—is emblematic of the transformational power of industrial modernity. Western mining communities formed in the late nineteenth and early twentieth centuries were among the most technologically advanced places on the planet. Butte, Montana, the remote copper capital, in the early 1880s became one of the first American cities to have electric lighting. The Coeur d'Alenes went from having no railroad lines in 1885, when Noah Kellogg's jackass made its propitious discovery, to having two by 1891. In 1903 the Washington Water Power Company completed a sixty-thousand-volt electrical power line from Spokane Falls, Washington, to Burke, Idaho, a mining and milling center in the Coeur d'Alenes. At eighty miles, it was the longest high-voltage power line in the world.[15]

In the late nineteenth and early twentieth centuries, the Coeur d'Alenes witnessed a rapid influx of workers from around the globe, the attraction of a number of the nation's foremost capitalists and mining engineers, and the rapid establishment of several large mining corporations. The mining district represented a sort of idealized version of rapid modernization.[16] The Coeur d'Alenes' intense and violent labor strife of the 1890s, however, serves to leaven any tendency toward bland celebration of what often has been a very raw, if not brutal, process.[17] Following the labor violence, the Bunker Hill Company decided to forge the town of Kellogg into its antiunion Gibraltar. In keeping with "progressive" corporate labor practices of the time, Bunker undertook numerous forms of corporate welfare and transformed Kellogg into a type of company town. Hewing to this paternalistic approach, in 1911 the company built and helped maintain an "industrial YMCA" in the heart of uptown Kellogg. There the Wardner Industrial Union, a company-sponsored union intended to preempt renewed efforts at independent unionization, was granted a floor for its own use.

If one face of modernity is rapid industrial and urban development—examples abound, ranging from "resource frontiers" such as the recent oil boom towns of western North Dakota and eastern Montana to the rise of megacities such as Chicago during the second half of the nineteenth century and Shanghai over the past few decades—another is accelerated entropy and decay. While major financial and governmental centers such as New York City and Brussels may be relatively well insulated against the effects of rapid economic downturns, industrial, resource-extractive, and tourist communities readily can be hollowed out in very little time. Since the deindustrialization of the United States commenced in earnest in the 1970s, numerous American towns and cities have lost their economic rationale virtually overnight and have suddenly found themselves scrambling to develop a new modus operandi. Steel towns like Youngstown, Ohio, textile enclaves such as Mount Airy, North Carolina, lumber communities like Potlatch, Idaho, and mining and smelting cities such as Butte and Kellogg—all have faced the difficulty of suddenly lost jobs, wages, consumer spending, tax bases, and community identities. The "Great Recession" of recent years demonstrated that even high-flying recreational communities such as Las Vegas and Orlando are also vulnerable to the effects of economic downturns.

One of the hallmarks of modernity is the greatly increased pace of social change over previous epochs. The rule is change, not stasis. This applies with especial force to global capitalism, a dynamic system that is marked by often unpredictable and irregular flows of commodities, workers, and capital. In 1981–1982, following the frantic but ultimately unsuccessful efforts of Coeur d'Alenes residents to find a buyer for Bunker Hill, the tide of the region's century-old economy swiftly receded. Dazed miners, smelter workers, merchants, and local government officials were left to pick up the pieces as they watched all that had appeared solid in their world suddenly melt into air.[18] By the mid-1980s Kellogg's civic leaders had made the fateful decision to transform the gritty mining and smelting town into a Bavarian-themed ski resort community. Raising the capital to build a gondola connecting the town to the nearby ski hill and enacting a city ordinance requiring merchants to erect signage in "Bavarian" lettering, while hardly easy, ultimately proved small challenges as compared with the difficulty of developing a successful resort community and forging an authentic new identity for the old mining and smelting town. The difficulties that Kellogg and the rest of the Coeur d'Alenes have faced in successfully reinventing themselves in a largely postindustrial setting are reflective of the sorts of issues faced by many of the deindustrialized communities that presently dot the United States and the world.

It is hoped that there is something instructive in this study of the environmental and social history of a small, albeit immensely profitable and polluted, mining and smelting region. Although perhaps somewhat extreme—few places have yielded comparable quantities of metals or levels of pollution—the story of the Coeur d'Alenes contains many of the elements that have decisively molded the shape of the United States and its communities over the past century and a half or so. These include the overarching quest for profits and power on the part of much of its social and economic elite; the sharply alienating and objectifying relationship with the natural world that characterized the nineteenth and much of the twentieth century; the advent in the latter part of the twentieth century of a new environmental view in the developed world that places greater value on ecological relationships and on the intrinsic qualities of nonhuman nature; and the tendency of business corporations to shift social costs like pollution onto the environment and workers and then to leave the bill for society at large to pay. Superficially, the Coeur d'Alenes may not resemble much of the rest of the Unites States, or the world, but the forces of modernity that molded (and are still molding) its social and environmental world remain at work all around us.

EPILOGUE

Changqing, Shaanxi Province, China: In August 2009, 851 children living near the Shaanxi lead and zinc smelter in north-central China tested positive for lead poisoning. More than 170 were hospitalized. Angry residents of Changqing and a nearby village marched on the Shaanxi smelter, the fourth-largest lead smelter in China, tearing down fences and attacking trucks before police restored order.[1] In response, the Chinese government closed the smelter. Although the Chinese environment minister, Zhou Shengxian, called for "powerful measures" to prevent pollution by lead and other heavy metals, many were skeptical that any real change would come. Enforcement of national regulations was problematic in China, where local officials often emphasized short-term profits over the health effects of industrial pollution.[2]

The lead poisoning episode at Changqing was not the only one to occur in August 2009. Days later a case involving 1,300 children was reported at an illegal manganese smelter operating in Hunan Province. Later in the month two hundred children in Kunming, the capital of Yunnan Province, showed signs of lead poisoning during a routine testing of one thousand children.[3] Sadly, the lead poisoning of children is not uncommon in China's many smelter towns. A Chinese health official estimates that between 50 and 60 percent of children under fourteen in the country's mining regions suffer lead poisoning.[4] Such regions are numerous in a nation that has emerged as the world's largest producer of refined lead and lead batteries. China's lead output rose nearly 20 percent in 2008, to 3.26 million tons.[5]

The export of dirty industries from wealthier to poorer regions of the globe, a hallmark of contemporary capitalism, is a process that shows no sign of abating. In 1977, for example, there were six primary lead smelters operating in the United States.[6] By 1995 only two remained. Now there are none. The Doe Run facility at Herculaneum, Missouri, thirty-five miles south of Saint Louis, closed for good in 2013. Primary lead smelting has arisen in places like China where development pressures are strong and environmental regulations lax. With the recent outcry over lead poisoning cases and the closing of some lead plants, however, there has been a trend for smelters to move to poorer, more remote areas of China.[7]

While American parents and children no longer need fear a smelter-caused lead intoxication epidemic of the kind that gripped the Kellogg area in the 1970s, China's outbreaks of childhood lead poisoning in smelter communities seem all too familiar. They belong to a larger history that is as old as the ancient Greeks and as new as the latest lead smelter to be assembled in a poor country.

NOTES

Introduction

1. I refer to "the Bunker Hill Company" here, but the company, founded in 1887, had been a wholly owned subsidiary of Houston-based Gulf Resources and Chemical Company since June 1, 1968.

2. Cassandra Tate, "Bunker Hill Agrees to Join State Study," *Lewiston Morning Tribune*, October 6, 1974, 1.

3. Tate, "Bunker Hill Agrees," 1. The blood lead screenings were conducted in August 1974 by Idaho's Department of Health and Welfare and the federal Centers for Disease Control. Although many suspected Bunker's culpability at the time, it had not yet been proven that the company's decision to run its lead smelter at full bore for months following the September 1973 baghouse fire that wrecked the smelter's emissions filtration system was the cause of the lead health epidemic.

4. ILZRO's previous study of a similar, albeit less severe situation involving a community nearby a smelter at El Paso, Texas, contradicted the conclusions reached by public health officials. The study minimized health effects and shielded the smelter owner, ASARCO, from blame. In a study of Toronto lead smelter workers that was widely ridiculed in scientific circles, ILZRO concluded that the smelter workers' lead exposures were benign since they lived twenty years longer than non-smelter workers.

5. "Industry-Funded Research of Lead Gets under Way," *Spokane Daily Chronicle*, September 28, 1974, 3.

6. Andrus is something of a legend in Idaho political history. The only governor to be elected to four terms in Idaho history, Andrus left office during his second term to serve as President Jimmy Carter's secretary of the interior.

7. Memorandum Re: Gulf Resources & Chemical Corporation—ILZRO, 2 October 1974, Manuscript Group (herafter MG) 413, Box 243, Folder 4455, Bunker Hill Mining Company Records, Special Collections, University of Idaho Library (hereafter Bunker Hill Records). In a high-level meeting between Gulf-Bunker executives and

ILZRO's top brass during the first days of October 1974, a Bunker official mentioned that the Kellogg lead epidemic had caused "great political harangue in Idaho," from which company officials evidently were feeling some heat.

8. "Public Hearing," 3 March 1971, p. 26, MG 413, Box 242, Folder 4547, Bunker Hill Records.

9. Nonetheless, the Kellogg City Council unanimously supported Gulf-Bunker's request for an extension of time to comply with Idaho's sulfur dioxide standards.

10. Ulrich Beck, *Risk Society: Towards a New Modernity* (London: Sage, 1992), 13.

11. Beck, *Risk Society*, 10, 13, 49.

12. A note on terminology: although the northern Idaho locale that is the subject of this work is today more commonly referred to as the Silver Valley than as the Coeur d'Alenes, in this work I employ the latter term more commonly than the former. I make this choice primarily because the Coeur d'Alenes is what the region was called for much of its history. The Silver Valley is a relatively recent appellation, dating to the early 1970s. In keeping with contemporaneous usage, I use "the Silver Valley" more commonly in discussing the last few decades in the area's history and "the Coeur d'Alenes" when dealing with earlier eras.

13. The Coeur d'Alene basin includes the North and South Forks of the Coeur d'Alene River, their tributaries, the Coeur d'Alene River, Lake Coeur d'Alene, and the Spokane River.

14. I include both human and nonhuman nature in my definition of *environment*. Human health, therefore, is very much an environmental concern of this work.

15. Matthew Klingle, *Emerald City: An Environmental History of Seattle* (New Haven, CT: Yale University Press, 2007), 4.

16. Klingle, *Emerald City*, 11.

17. Emory Jerry Jessee, "Radiation Ecologies: Bombs, Bodies, and Environment during the Atmospheric Nuclear Weapons Testing Period in the United States, 1942–1965" (PhD diss., Montana State University–Bozeman, 2013). The seminal work in this regard is of course Rachel Carson's *Silent Spring*. Published in 1962, the book introduced many to the idea of environmental threat from pesticides such as DDT.

18. Timothy J. LeCain, *Mass Destruction: The Men and Giant Mines That Wired America and Scarred the Planet* (New Brunswick, NJ: Rutgers University Press, 2009). The notion that the base materials that form essential elements in modern consumer products (e.g., copper in car engines) and the often environmentally destructive practices that produce them for the market are not well understood by the consuming public is central to LeCain's work. Since consumers are generally unaware of the connection between, for instance, open-pit copper mining and their computers or automobiles, they are unlikely to see the inherent environmental tradeoffs involved in their

purchases, and, more generally, in much of modern consumer society. My own work owes a substantial debt to LeCain's scholarship.

19. Ulrich Beck, *World at Risk* (Cambridge, UK: Polity, 2009), 22.

20. Marshall Berman, *All That Is Solid Melts into Air: The Experience of Modernity* (New York: Simon and Schuster, 1982), 294. New York City's master developer Robert Moses, though a fervid anticommunist, also liked to employ this phrase.

21. Berman, *All That Is Solid*, 68.

22. Berman, *All That Is Solid*, 78, 98, 102.

23. From 1885 to 1997, in 1997 dollars, the most recent statistic available to the author.

24. A relatively handful of high-paying mining jobs still exist in the district.

Chapter 1: Uncle Bunker's Town

1. Murray lay at the center of the 1883–1884 Prichard Creek Gold rush. By 1886, however, the rush was petering out.

2. Katherine G. Aiken, *Idaho's Bunker Hill: The Rise and Fall of a Great Mining Company, 1885–1981* (Norman: University of Oklahoma Press, 2005), 17.

3. Frederick Burbidge to John A. Finch, 17 December 1900, MG 367, Box 9, Folder 89; and Stanly Easton to F. W. Bradley, 1 January 1907, Box 10, Folder 109, Bunker Hill Records. The letterhead of 1900 listed "Mines and Works at Wardner, Idaho," and "Postoffice address Kellogg, Idaho," while the only geographic reference in the letterhead of 1907 was to "Kellogg, Idaho."

4. Pullman Palace Car Company employees were required to live in Pullman, south of Chicago, and purchase their supplies from the company store. The town of Potlatch, located about ninety miles southwest of Kellogg as the crow flies, was founded by the Potlatch Lumber Company, a subsidiary of the Weyerhauser Corporation, in 1907. The town was created to house workers at the company's newly constructed white pine sawmill, at the time the world's largest. In both Pullman and Potlatch, all of the real estate was owned by the respective companies. In addition, by company decree both towns were "dry."

5. John S. Garner, ed., *The Company Town: Architecture and Society in the Early Industrial Age* (New York: Oxford University Press, 1992), 3.

6. Garner, *Company Town*, 8.

7. Kellogg undoubtedly was more of a company town during the 1903–1942 period, when Bunker Hill workers were represented only by the Wardner Industrial Union, a company union, than it would be in 1942–1982, when company workers were represented by, over time, the International Union of Mine, Mill and Smelter Workers; the Northwest Metal Workers Association; and the United Steelworkers of America.

During this latter period, the union served as a counterweight to company authority and identity. The union hall became a community center to rival the Bunker Hill–sponsored YMCA and Staff House. As signs of these shifts, in the 1950s the annual Bunker-sponsored Miners' Picnic died out and the company stepped back from its longtime role as a housing provider in Kellogg.

8. Dwight Jensen, "The Loneliness of the Environmental Reporter," *Columbia Journalism Review* 15, no. 5 (January–February 1977): 40–42. Brainerd succeeded his father as editor of the newspaper, which served as Kellogg's only daily for many decades. In the article, Bunker president James Halley also revealed that the *Evening News* published basically anything the company sent it to print.

9. US Census Office, "Table 5—Population of States and Territories by Minor Civil Divisions: 1890 and 1900," in *Census Reports*, vol. I, *Twelfth Census of the United States Taken in the Years 1900: Minor Civil Divisions* (Washington, DC: Government Printing Office), 113.

10. US Census Office, *Thirteenth Census of the United States Taken in the Year 1910*, vol. 2, *Population Reports by States with Statistics for Counties, Cities and Other Civil Divisions, Alabama-Montana, Population—Idaho* (Washington, DC: Government Printing Office), 418–19, table 1.

11. "Kellogg Population Estimated at 5,900," *Lewiston Morning Tribune*, November 9, 1957, 10.

12. US Bureau of the Census, "Chapter A, Number of Inhabitants, Part 14, Idaho: Population of County Subdivisions: 1960–1980," in *1980 Census of Population*, vol. 2, *Characteristics of the Population* (Washington, DC: Government Printing Office), 12.

13. David Bender, "Tiny Idaho Cities Make Big Splash in Census," *Spokesman-Review*, February 3, 1991. The most recent estimate of Kellogg's population, ca. 2009, is 2,177. See http://idaho.hometownlocator.com/id/shoshone/kellogg.cfm (accessed August 16, 2011).

14. US Census Office, *Thirteenth Census of the United States Taken in the Year 1910*, vol. 2, 587, table 21.

15. US Census Office, *Thirteenth Census of the United States Taken in the Year 1910*, vol. 2, 587, table 21.

16. The highest, albeit temporary, populations of African Americans were likely achieved when an all-black army unit was stationed in Shoshone County following the labor violence of 1899.

17. US Census Office, *Compendium of the Eleventh Census: 1890*, part 1, *Population* (Washington, DC: Government Printing Office), 481, table 13; US Bureau of the Census, *15th Census of the United States: 1930, Population*, vol. 3, part 1, *Reports by States,*

Showing the Composition and Characteristics for Counties, Cities, and Townships or Other Minor Subdivisions, Alabama-Missouri (Washington, DC, 1932), 565, table 11.

18. Julie Whitesel Weston, *The Good Times Are All Gone Now: Life, Death, and Rebirth in an Idaho Mining Town* (Norman: University of Oklahoma Press, 2009), 31, 33.

19. Bunker Hill's corporate headquarters were located in San Francisco and Bradley split his time between "the City by the Bay" and Kellogg.

20. Aiken, *Idaho's Bunker Hill*, 48.

21. Aiken, *Idaho's Bunker Hill*, 48. In his confession, Orchard also admitted to having planted the bomb that killed Steunenberg in 1905. That bomb was attached to the yard gate of the former governor's home in Nampa, Idaho.

22. Aiken, *Idaho's Bunker Hill*, 12.

23. Aiken, *Idaho's Bunker Hill*, 13.

24. Aiken, *Idaho's Bunker Hill*, 21, 25.

25. Aiken, *Idaho's Bunker Hill*, 25–26.

26. Aiken, *Idaho's Bunker Hill*, 27.

27. Aiken, *Idaho's Bunker Hill*, 32.

28. Stanly Easton, like Bradley a UC, Berkeley–trained mining engineer (class of 1895), came to work for Bunker in 1902 as assistant mine manager. In 1903, Easton was promoted to mine manager, a position he would hold until Bradley's death in 1933. In that year, Easton assumed the company presidency. Easton served as Bunker Hill president until 1954 and as the company's chairman of the board of directors from 1954 to 1958, when he retired.

29. Aiken, *Idaho's Bunker Hill*, 32.

30. Easton to Bradley, 13 September 1912, 28 September 1912, and 12 December 1912, all MG 367, Box 10, Folder 119, Bunker Hill Records.

31. Aiken, *Idaho's Bunker Hill*, 12, 51; Easton to Bradley, 13 May 1906, and Bradley to Easton, 18 May 1906, MG 367, Box 9, Folder 107, Bunker Hill Records.

32. Andrea Tone, *The Business of Benevolence: Industrial Paternalism in Progressive America* (Ithaca, NY: Cornell University Press, 1997), 52, 42.

33. Tone, *Business of Benevolence*, 52, 54. Interestingly, the move toward "corporate welfare" occurred nearly a century earlier in the British coal country, where by the 1830s "coal owners increasingly viewed education, religion, and benefit societies as important means of securing a stable, as well as a deferential workforce." James Alan Jaffe, *The Struggle for Market Power: Industrial Relations in the British Coal Industry, 1800–1840* (Cambridge, UK: Cambridge University Press, 1991), 89.

34. Tone, *Business of Benevolence*, 33, 35–37.

35. Tone, *Business of* Benevolence, 65.

36. Tone, *Business of Benevolence,* 241.

37. Bradley to Frederick Burbidge, 18 May 1900, p. 1, MG 367, Box 9, Folder 104, Bunker Hill Records.

38. Burbidge to F. W. Bradley, 26 September 1899, MG 367, Box 8, Folder 86, Bunker Hill Records.

39. Easton to C. W. Evans [financial secretary, Wardner Industrial Union], 6 October 1905, MG 367, Box 9, Folder 106, Bunker Hill Records.

40. Aiken, *Idaho's Bunker Hill,* 80–81, 106–7. Under the pressure of New Deal support for unionization and the popularity of the International Union of Mine, Mill and Smelter Workers in the mid-1930s, Bunker allowed the WIU to form an Employee Representation Committee to meet with company officials and make formal requests. Bunker denied the committee's more costly requests for more pay and paid vacations, while honoring some that were less expensive to meet.

41. Bradley to Easton, 17 March 1911, MG 367, Box 10, Folder 117, Bunker Hill Records.

42. Ray Chapman, *Uncle Bunker: Memories in Words and Pictures* (Kellogg, ID: Chapman, 1994), 24–25.

43. "Minutes of Meeting—Trustees—Hecla-Bunker Hill Scholarship Fund," 27 April 1977, MG 367, Box 14, Folder 224, Bunker Hill Records. From 1952 to 1977, the scholarship fund donated $165,399.41 to local students. Until a 1970s Internal Revenue Service ruling sidelined the practice, preference was given to local students who either were themselves current or former Bunker employees or whose parents had that status.

44. Easton to Bradley, 25 April 1917, and Bradley to Easton, 28 April 1917, MG 367, Box 10, Folder 128, Bunker Hill Records.

45. Easton to Bradley, 25 April 1917; and Bradley to Easton, 28 April 1917.

46. Ray Chapman, *History of Kellogg, Idaho: 1885–2002* (Kellogg, ID: Chapman, 2002), 12. The YMCA building was shuttered in 1979 when, due to fire and disability access regulations, it became cost-prohibitive to maintain.

47. Bradley to Easton, 1 February 1910, MG 367, Box 10, Folder 115, Bunker Hill Records.

48. Bradley to Easton, 15 February 1910; MG 367, Box 10, Folder 115, Bunker Hill Records.

49. R. D. Wilder, "Kellogg's Industrial Y.M.C.A.," *Engineering and Mining Journal,* August 1939, 81.

50. The company collected rent from the Wardner Industrial Union, but this amount did not come close to covering its expenses for the facility.

51. Easton to J. S. Wallace [F. W. Bradley's personal assistant], 20 October 1917, MG 367, Box 10, Folder 130, Bunker Hill Records. In the letter, Easton pointed out that the YMCA's "overdue bill" of $434.75 "represents a donation made by this Company, to the association." "Bunker Hill & Sullivan M & C Co Kellogg Salaries," MG 367, Box 10, Folder 126, Bunker Hill Company Records, lists the salary of J. E. Sturtevant, the YMCA director, as $125 for January 1916 and $155 for November 1916.

52. Easton to Bradley, 1 October 1917, MG 367, Box 10, Folder 130, Bunker Hill Records.

53. Easton to Bradley, 1 October 1917. Bradley to Easton, 5 October 1917, Bunker Hill Records.

54. Ibid., Easton to Bradley, 21 December 1917. Weston, *Good Times Are All Gone Now*, 147.

55. Easton to Bradley, received 21 October 1924, MG 367, Box 12, Folder 149, Bunker Hill Records.

56. Wilder, "Kellogg's Industrial Y.M.C.A.," 82.

57. Wilder, "Kellogg's Industrial Y.M.C.A.," 82.

58. Thomas Winter, *Making Men, Making Class: The YMCA and Workingmen, 1877–1920* (Chicago: University of Chicago Press, 2002), 30.

59. Winter, *Making Men*, 33, 90, 39.

60. Winter, *Making Men*, 7.

61. Winter, *Making Men*, 65.

62. Winter, *Making Men*, 12, 61.

63. Winter, *Making Men*, 14.

64. Winter, *Making Men*, 147.

65. Thomas Arthur Rickard, *The Bunker Hill Enterprise: An Account of the History, Development, and Technical Operations of the Bunker Hill & Sullivan Mining & Concentrating Company at Kellogg, Idaho, U.S.A.* (San Francisco: Mining and Scientific Press, 1921), 143.

66. "Bunker Hill and District Recalled in Feature," *Kellogg Evening News*, July 1, 1974.

67. Jack Bradley was the youngest son of former company president F. W. Bradley. Married to Jane Easton, Stanly Easton's daughter, the younger Bradley was, like his father and father-in-law, a mining engineer trained at the University of California, Berkeley. The ascension of the forty-five-year-old Bradley to Bunker's helm represented a generational shift in the company's management. Although under Bradley's leadership Bunker pulled back from involvement in the Kellogg housing market, the company never completely withdrew from that field. Housing shortages in the area drew Bunker back into housing development in 1968, when the company became involved in a housing

project on the 160-acre parcel where the old Kellogg Airport had been, and again in the mid-1970s when it joined with other area mining projects in developing a worker housing project in Osburn.

68. "Bunker Announces Important Changes," *Kellogg Evening News*, June 12, 1956, 1; "Management Committee Meeting," 9 March 1956, MG 367, Box 6, Folder 40, Bunker Hill Records.

69. "Bunker Announces Important Changes," 1.

70. "Bunker Announces Important Changes"; "WWP Purchase of System OK'd," *Spokesman-Review*, August 24, 1956, 1. In addition, Bunker operated a water business for its own operations that also provided domestic service to some residents of Kellogg, Wardner, and Smelterville. The company provided this service from the early years of the century until October 1974, when it sold its domestic water business to the Central Shoshone County Water District for $85,000.

71. "Bunker Announces Important Changes," 1.

72. Easton to Bradley, 1 November 1907, MG 367, Box 10, Folder 110, Bunker Hill Records.

73. Easton to Bradley, 1 November 1907.

74. Easton to Bradley, 13 May 1913, MG 367, Box 10, Folder 121, Bunker Hill Records.

75. Aiken, *Idaho's Bunker Hill*, 54.

76. Rickard, *Bunker Hill Enterprise*, 141.

77. Rickard, *Bunker Hill Enterprise*, 140. Bunker let the mine become part of Kellogg in 1955 and the lead smelter in 1956.

78. Easton to Abbott, 30 September 1913, MG 367, Box 10, Folder 119, Bunker Hill Records.

79. Rickard, *Bunker Hill Enterprise*, 141.

80. "Liberty Loan Bonds," attachment to letter from Easton to Bradley, 5 June 1917, MG 367, Box 10, Folder 128, Bunker Hill Records.

81. Easton to Bradley, 22 June 1917, MG 367, Box 10, Folder 128, Bunker Hill Records.

82. "Notice to Employees," 23 July 1917, MG 367, Box 10, Folder 129, Bunker Hill Records.

83. A common nickname for IWW members was "Wobblies."

84. Bradley to Easton, 16 July 1917, MG 367, Box 10, Folder 129, Bunker Hill Records.

85. James W. Byrkit, "The IWW in Wartime Arizona," *Journal of Arizona History* 18, no. 2 (1977): 149–70.

86. George Everett, "Frank Little's Grave," *Only in Butte*, accessed November 8, 2011, http://www.butteamerica.com/flit.htm.

87. Keith C. Petersen, *Company Town: Potlatch, Idaho, and the Potlatch Lumber Company* (Pullman: Washington State University Press, 1987), 164.

88. Easton to Bradley, received 22 March 1917, MG 367, Box 10, Folder 127, Bunker Hill Records.

89. Easton to Bradley, received 22 March 1917; Easton to Bradley, 18 April 1917.

90. Easton to Bradley, received 2 July 1917, MG 367, Box 10, Folder 128, Bunker Hill Records.

91. Easton to Bradley, received 2 July 1917.

92. Easton to Bradley, 8 July 1917, MG 367, Box 10, Folder 129, Bunker Hill Records. The quotation from Mayor Mason is from an editorial titled "Kellogg Shows the Way." Easton included the editorial, drawn from an undetermined Wallace newspaper, with his letter to Bradley. As of the end of August 1917, roughly 180 men had been sworn in as special policemen by the City of Kellogg and forty-six Bunker Hill employees had been deputized by the Shoshone County sheriff.

93. Easton to Bradley, 8 July 1917.

94. Easton to Bradley, 8 July 1917; Bradley to Easton, 16 July 1917.

95. Section 7(a) of the National Industrial Recovery Act, signed into law by President Franklin Roosevelt in July 1933, permitted workers to bargain collectively. The federal law spurred a rash of union organizing around the country, including the Coeur d'Alenes. The vote of May 26, 1942, to unionize was quite close, 353 for Mine-Mill representation and 327 against.

96. Aiken, *Idaho's Bunker Hill*, 132.

97. Rickard, *Bunker Hill Enterprise*, 141.

98. Easton to Bradley, 9 January 1925, MG 367, Box 12, Folder 150, Bunker Hill Records.

99. Easton to Bradley, 9 January 1925; Easton to Bradley, 20 February 1925. Easton also puzzled over how to handle situations regarding two widows of "former good employees" (only current, full-time employees were supposed to be eligible to take advantage of the offering). Before deferring the matter to another day, Easton informed Bradley that "these cases are wholly worthy, the women are permanent residents of Kellogg, are properly bringing up and educating good sized families and are husbanding their meager resources and making a brave effort to add to them." In another case, that of a college student, "the son of an employe[e] who works here during vacation and . . . will without doubt in due course follow his father's experience and be a regular employe[e] of the Company," Easton made an exception, allowing him to purchase the stock.

100. Easton to Bradley, 20 February 1925; Easton to Bradley, 1 December 1925, MG 367, Box 12, Folder 153, Bunker Hill Records. The second offering of the pre-

ferred stock was made between February and December 1925. The par value of the stock was approximately $200,000. It is not clear whether the stock plan was continued the following year. The plan was revived later, however. An additional 1,010 shares of preferred stock were issued for employee purchase in 1951. See Board of Directors Meeting Minutes, 17 November 1951 meeting, MG 367, Box 2, Folder 8, Bunker Hill Records.

101. The "company union" was replaced by the decidedly more independent, aggressive International Union of Mine Mill and Smelter Workers (IUMMSW) in late 1943.

102. "Ten Thousand Persons Gather at Kellogg Today for Annual Miners' Picnic," *Spokane Daily Chronicle*, 3.

103. The latter in honor of the legendary burro reputed to have led Noah Kellogg to the silver-lead outcroppings that in turn led to the development of the mining district.

104. "Kellogg Fete Dates Are Fixed," *Spokesman-Review*, July 27, 1916, 10.

105. "Scholarships Granted to Thirty-Seven," *Bunker Hill Reporter*, June 1968, p. 1, MG 367, Box 83, Folder 1868, Bunker Hill Records. A handful of scholarships was awarded to nonlocals studying mechanical or mining engineering, or related fields, at Idaho institutions of higher learning.

106. "Students Notified Applications for Scholarships Due," *Bunker Hill Reporter*, March 1958, p. 1, MG 367, Box 83, Folder 1869.5, Bunker Hill Records.

107. "Minutes of Meeting—Trustees: Hecla-Bunker Hill Scholarship Fund," 27 April 1977, MG 367, Box 14, Folder 224, Bunker Hill Records.

108. "Jobs at Bunker Aid Local Youth," *Bunker Hill Reporter*, August 1959, p. 2, MG 367, Box 83, Folder 1870, Bunker Hill Records.

109. The United Crusade later became the United Way.

110. *Bunker Hill Reporter*, April 1958, p. 3, MG 367, Box 83, Folder 1869.5, Bunker Hill Records.

111. Weston, *Good Times Are All Gone Now*, 145.

112. Directors Meeting Minutes, 27 June 1957. *Bunker Hill Reporter*, September 1962, p. 1, MG 367, Box 83, Bunker Hill Records.

113. "Boy Scouts Consider Selling Camp Easton to Developer," *Spokesman-Review*, August 13, 2011, accessed September 29, 2011, http://www.spokesman.com/stories/2011/aug/13/boy-scouts-consider-selling-c-easton-to/.

114. "Scout Masters to Get Leave," *Bunker Hill Reporter*, p. 4, MG 367, Box 83, Folder 1871, Bunker Hill Records.

115. "Bunker Plays Active Role in XII World Jamboree," *Bunker Hill Reporter*, August 1967, p. 3, MG 367, Box 83, Folder 1877, Bunker Hill Records.

116. "Kellogg YMCA Is Now 50 Years Old," *Bunker Hill Reporter,* September 1961, p. 2, MG 367, Box 83, Folder 1871, Bunker Hill Records.

117. "Kellogg YMCA Is Now 50 Years Old," 2. "Kellogg YMCA Night School Is 'In Business,'" *Bunker Hill Reporter,* December 1966, p. 3, MG 367, Box 83, Folder 1876, Bunker Hill Records.

118. Chapman, *History of Kellogg,* 12.

119. "Bunker Hill 'Kids,' Adults Participate [in] YMCA Programs," *Bunker Hill Reporter,* February 1966, p. 4, MG 367, Box 83, Folder 1876, Bunker Hill Records. "Simmons Is Re-Elected Head of YMCA Board," *Bunker Hill Reporter,* March 1959, p. 1, MG 367, Box 83, Folder 1870, Bunker Hill Records. The *Bunker Hill Reporter* was the company's staff publication. In addition to its support for the Y, Bunker also sponsored a winter bowling league and granted it ample coverage in the *Reporter.*

120. Unidentified, undated document, MG 367, Box 72, Folder 1511, Bunker Hill Records. This appears to be an internal company document prepared by Bunker's negotiations team circa early 1952.

121. Among the successful mining companies in the district in the 1950s were the Sunshine Mining Company, owner of the Sunshine Mine, the world's greatest silver mine, located between Kellogg and Osburn, and the Hecla Mining Company, owner of the Lucky Friday Mine, located near Mullan.

122. "Kellogg Easily Tops Business: 1954 Business Census Figures Revealed," *Kellogg Evening News,* February 6, 1955, 1. Kellogg trailed its rival, Wallace, the county seat, in the categories of general merchandise stores; apparel and accessory stores; and furniture, home furnishings, and appliance stores. Although Kellogg and Wallace each had three drugstores, sales from Kellogg's were by far the greater of the two.

123. "Shoshone County Families Earn More than Average in Idaho," *Kellogg Evening News,* January 11, 1957, 3. The business year runs from July 1 to June 30.

124. While the closure of Bunker's operations, and the layoff of 2,100 workers, in February 1982 undoubtedly constituted the greatest blow to the area's economy, the June shutdown of the six-hundred-employee Sunshine Mine, the world's most productive silver mine, was another major hit. Unlike Bunker, however, Sunshine eventually returned, albeit sporadically, to full operations.

125. *GAP News,* 3 March 1976, MG 367, Box 83, Folder 1867, Bunker Hill Records.

126. Unidentified, undated document, probably from early 1952, MG 367, Box 72, Folder 1511, Bunker Hill Records. "Two Buyers Negotiating for Bunker," *Spokesman-Review,* September 16, 1981, 8. The employment low occurred in 1944, during the World War II manpower shortage, and the employment high in 1956. The highest payroll was made during the silver price spike year of 1980 and the lowest payroll came in 1945, the final full year of WWII rationing. The payroll numbers are not adjusted

for inflation; however, if put into 1942 dollars, the $50 million 1980 payroll becomes $8,976,660. The approximate averages for the period are two thousand workers and a $16 million payroll.

127. "C. E. Schwab Talk before Kellogg Chamber of Commerce," 26 January 1961, MG 367, Box 83, Folder 1858, Bunker Hill Records.

128. A. Y. Bethune, "What the Bunker Hill Means to Spokane and the Inland Empire," talk before the Spokane North Side Rotary Club, 22 May 1961, p. 4, MG 367, Box 83, Folder 1858, Bunker Hill Records. Bunker long held the status of both the Washington Water Power Company's and the Union Pacific's largest customer in north Idaho.

129. Bethune, "What the Bunker Hill Means." The value of that $47 million in 2011 dollars would be $345,450,000.

130. "Bunker Hill Stock Transfer Procedure Is Interesting Part of Corporate Affairs," *Bunker Hill Reporter,* September 1962, p. 2, MG 367, Box 83, Folder 1872, Bunker Hill Records. In addition, under Gulf Resources, the Houston-based company that absorbed Bunker Hill in 1968, Bunker's operations were headquartered in Spokane from 1968 to 1971, before being returned to Kellogg. Bunker, over the years, also held some of its shareholder meetings in Spokane.

131. "Remarks Prepared by John D. Bradley for Presentation before Northwest Mining Association," p. 2, MG 367, Box 83, Folder 1858, Bunker Hill Records. The talk was delivered by Wallace G. Woolf on December 4, 1959, following Bradley's untimely death in an auto accident the previous Thanksgiving Day. The value of the $19 million referenced by Bradley in 2011 dollars would be $139,650,000, "North Idaho Towns Have Top Tax Rates," *Spokane Daily Chronicle,* December 24, 1957, 3. The average per capita property tax rate in Kellogg for 1957 was $117.19, although few, if any, individual Kellogg homeowners paid that much.

132. "Property Taxes Up," *Bunker Hill Reporter,* November–December 1975, p. 1, MG 367, Box 83, Folder 1885, Bunker Hill Records.

133. Before this time, the company's stock was traded on regional exchanges such as the Spokane Stock Exchange.

134. Schwab was soon thereafter hired to a senior management position by the Butte, Montana, copper giant, the Anaconda Company.

135. "Minutes of Meeting—Trustees: Hecla-Bunker Hill Scholarship Fund," 27 April 1977, MG 367, Box 14, Folder 224, Bunker Hill Records. "Telecast Order" from Spokane's KHQ-6TV, MG 367, Box 14, Folder 240, Bunker Hill Records. Bunker president E. Viet Howard to Bob Douglas, Kellogg Chamber of Commerce President, and Terry Smith, Manager, Idaho First National Bank, 21 March 1978, MG 367, Box 14, Folder 240, Bunker Hill Records.

136. "Confidential" report to senior Gulf officials by public relations firm Dale Henderson, Inc., 1975, p. 1, MG 413, Box 242, Folder 4545, Bunker Hill Records.

Chapter 2: The Useful Metal

1. United States Geological Service (USGS), "Lead Statistics and Information," accessed April 20, 2011, http://minerals.usgs.gov/minerals/pubs/commodity/lead/.

2. USGS, "Lead Statistics and Information." An additional use of lead was as an agricultural insecticide. Developed in 1893, lead arsenate coupled two toxic minerals, lead and arsenic, in a liquid compound that was effective in killing the gypsy moth and other insects. Although it was largely replaced by DDT, introduced in 1948, as late as 1947 approximately twenty-five thousand tons of lead arsenate was sold in the United States.

3. USGS, "Lead Statistics and Information." Bearing out these trends, US primary lead production went from 455,000 tons in 1913 to 744,000 tons in 1926.

4. Christian Warren, *Brush with Death: A Social History of Lead Poisoning* (Baltimore: Johns Hopkins University Press, 2000), 118. In the late 1950s and early 1960s, TEL accounted for roughly 20 percent of total US lead consumption, making it the second most substantial use of lead in the United States after storage batteries. By 1970, the US auto industry was consuming roughly 300,000 tons of TEL.

5. *Bunker Hill Reporter*, July–August 1973, p. 4, MG 413, Box 83, Folder 1883, Bunker Hill Records. Domestic demand for the metal remained remarkably strong and consistent in the postwar decades, dropping below a million tons only five times between 1947 and 1979, and falling beneath 900,000 tons only once.

6. Lead Industries Association, *Lead in Modern Industry* (New York: Lord Baltimore Press, 1951), 14–15.

7. Lead Industries Association, *Lead in Modern Industry*, 14–16.

8. Warren, *Brush with Death*, 18.

9. Warren, *Brush with Death*, 1.

10. Warren, *Brush with Death*, 18.

11. Warren, *Brush with Death*, 19–23.

12. Indeed, the words *plumber* and *plumbing* derive from *plumbum*, the Latin word for lead, as does a term denoting lead poisoning, *plumbism*.

13. Warren, *Brush with Death*, 19. To this day a common term for lead poisoning is *saturnism*. The Greeks considered lead the father of metals, as they viewed Cronus to be the father of the gods. In Egyptian mythology the god Osiris was murdered by being buried alive in a lead-sealed coffin. April McDevitt, "The Story of Isis and Osiris," *Ancient Egypt: The Mythology*, accessed October 24, 2011, http://www.egyptianmyths.net/mythisis.htm.

14. Wrist and foot drop refers to loss of the ability to hold one's wrist or foot up. This condition is prompted by lead-induced nerve damage. In the lead industries, wrist and foot drop have long been signs of the lead intoxication of workers.

15. New York State Department of Health, "Lead Exposure in Adults—A Guide for Health Care Providers," accessed May 17, 2011, http://www.health.state.ny.us/publications/2584/.

16. Warren, *Brush with Death,* 16; Lydia Denworth, *Toxic Truth: A Scientist, a Doctor, and the Battle over Lead* (Boston: Beacon, 2008), 26; Anna Marie Helmenstine, "What Makes Lead Poisonous," About.com, accessed October 25, 2011, http://chemistry.about.com/od/howthingsworkfaqs/f/leadpoisoning.htm. Heme is the iron-rich pigment without which blood cannot do its job of carrying oxygen throughout the body. Mitochondria are the parts of the cell structure responsible for producing energy.

17. Lead is generally more bioavailable to young children than to adults. Due to children's rapid growth rates and growing brains, lead poses a great threat to their health and their IQs. Because it crosses the placental barrier, lead represents a grave hazard to fetuses.

18. Warren, *Brush with Death,* 20–24.

19. Denworth, *Toxic Truth,* 29. Printers have history with lead that dates to the Renaissance. Following the invention of the printing press in the sixteenth century, movable type was made from lead-tin alloys. Later, linotype machines that automatically cast molten-lead type as an operator keyed in text to be printed subjected printers to the dangers of plumbism. Linotype machines continue to be used to this day.

20. Warren, *Brush with Death,* 68–69. Although Tanquerel's report boosted zinc-based paints and dealt a blow to France's lead paint industry, the same cannot be said for the United States, where white lead-based paints would retain their dominant position into the 1920s.

21. Hull House, a settlement house, was cofounded by Jane Addams in 1889 as a community of university women whose main purpose was to provide social and educational opportunities for working-class people (many of them recent European immigrants) in the surrounding neighborhood.

22. Warren, *Brush with Death,* 72.

23. Warren, *Brush with Death,* 76.

24. Warren, *Brush with Death,* 120. Lead oxides, of course, can be highly toxic as well, but generally only when breathed in or swallowed.

25. Warren, *Brush with Death,* 120. Still, the Chemical Warfare Service estimated that roughly five teaspoonfuls of TEL applied to healthy skin would prove fatal to an adult.

26. General Motors and Standard Oil of New Jersey together created the Ethyl Gasoline Corporation in August 1924. Ethyl held patent rights to the TEL gasoline additive, but did not manufacture the substance itself. It licensed the lion's share of the manufacture out to the DuPont Corporation and a small portion to Standard Oil of New Jersey. GM had experimented with the manufacture of TEL at its Dayton laboratories and originally tasked DuPont with the commercial manufacture of the substance, but soon became dissatisfied with DuPont's slow production rate. Standard Oil of New Jersey, with DuPont a major owner of GM, began experimenting with other methods of producing TEL and came up with a much faster, cheaper system. In order both to reward Standard and to benefit from its production methods, GM agreed to share TEL patent rights with Standard. This was accomplished via the creation of the Ethyl Gasoline Corporation as a joint venture of GM and Standard Oil of New Jersey. As part of the deal that created Ethyl, Standard of New Jersey was permitted to engage in commercial manufacture of a relatively small amount of TEL employing its new production methods. This Standard did, with fateful results, at its Bayway facility.

27. Jamie Lincoln Kitman, "The Secret History of Lead," *Nation*, March 20, 2000, 26.

28. David Rosner and Gerald Markowitz, eds., *Dying for Work: Workers' Safety and Health in Twentieth-Century America* (Bloomington: Indiana University Press, 1987), 125.

29. Warren, *Brush with Death*, 120.

30. Rosner and Markowitz, *Dying for Work*, 128. It is difficult to know exactly how many workers died and were sickened at Deepwater. DuPont only admitted that "some" workers had died before the Bayway incident. The company's tight control over worker information and over information flowing from the local hospital where employees were treated made uncovering information about worker health at Deepwater difficult. Reporters received much of their information from workers themselves.

31. Warren, *Brush with Death*, 120. Six more Deepwater workers died in early 1925 after DuPont began using Standard Oil's method of manufacturing TEL.

32. Warren, *Brush with Death*, 122.

33. Kitman, "Secret History of Lead," 26.

34. Rosner and Markowitz, *Dying for Work*, 129.

35. Kitman, "Secret History of Lead," 28. On May 4, in an act of brilliant public relations, Ethyl publicly withdrew its product from the market. TEL would remain off the market over a year.

36. Rosner and Markowitz, *Dying for Work*, 129.

37. Rosner and Markowitz, *Dying for Work*, 134–35.

38. Warren, *Brush with Death*, 128.

39. Rosner and Markowitz, *Dying for Work*, 135. The Ethyl-funded occupational lead science was spearheaded from the 1920s through the 1960s by Dr. Robert Kehoe at his research facility at the University of Cincinnati. Kehoe maintained that blood lead levels of up to 80 μg/dl could be safely borne by adults, and thus that atmospheric levels of TEL encountered in the United States were not harmful for the general population.

40. Warren, *Brush with Death*, 35–37.

41. Warren, *Brush with Death*, 33–34.

42. Warren, *Brush with Death*, 46. In addition, white lead-based paints were associated with the virtues of cleanliness, racial whiteness, and, ironically given the product's toxic nature, purity. Until it was supplanted by car batteries in the 1920s, lead paint constituted the largest destination of lead in the United States.

43. Warren, *Brush with Death*, 62, 202. Lead-based paints were not banned from interstate commerce in the United States until 1977.

44. Warren, *Brush with Death*, 142–43.

45. Warren, *Brush with Death*, 184–85.

46. Warren, *Brush with Death*, 169. BAL was developed secretly by the Allies during World War II as an anti-nerve-gas agent. It was tested successfully as a chelating agent for lead in late 1946. BAL works by binding with lead in the body to form a less harmful compound that the body can safely excrete. It is in no sense a "cure" for lead poisoning, but can substantially reduce blood lead levels in victims of lead poisoning. Blood lead levels usually drop substantially after a course of treatment with BAL, only to rise again later due to the fact that the chelating agent leaves a good deal of lead stored in the bone. That "bone lead" is prone to enter the bloodstream at a later time.

47. Warren, *Brush with Death*, 169.

48. Warren, *Brush with Death*, 202. The rule, promulgated by the Consumer Product Safety Commission, went into effect in 1978.

49. Denworth, *Toxic Truth*, ix.

50. Warren, *Brush with Death*, 211, 239, 261–62. While the average blood lead level for all Americans in 1965 was approximately 20 μg/dl, as of 1994 it was 2.8 μg/dl. Whereas between 1923 and 1966 an average of about twenty children per year died from lead poisoning, the most recent case of a child dying from lead poisoning in the United States appears to have occurred in 2006 when a four-year-old Minnesota boy swallowed a charm containing 99 percent lead. For more information on the Minnesota fatality, see Minnesota Department of Health, "Child Dies of Lead Poisoning from Metal Charm," accessed June 1, 2011, http://www.health.state.mn.us/divs/eh/lead/topics/braceletrecall.html.

51. Warren, *Brush with Death*, 229.

52. Denworth, *Toxic Truth*, 105.

53. Denworth, *Toxic Truth*, 213–14.

54. Kitman, "Secret History of Lead," 36.

55. Warren, *Brush with Death*, 220–21.

56. Denworth, *Toxic Truth*, 104. Before the mobilization of a contingent of antilead scientists including Herbert Needleman and Sergio Piomelli, the EPA had proposed a standard of 5.0 μg/dl, more than three times higher than the eventual standard.

57. Denworth, *Toxic Truth*, 202–3. By 2000, there had been eight major prospective studies on lead conducted around the world. Six of these demonstrated evidence of a harmful effect from lead at very low levels. Combining the data from seven of the eight studies—encompassing 1,333 children with a mean blood lead level of 12.4 μg/dl (slightly over the current standard of concern in the United States)—showed a decline of 6.2 IQ points for an increase of 10 μg of blood lead. Further, the steepest losses of IQ came at the lowest levels of lead exposure (e.g., at 1–7 μg/dl). The EPA's lead criteria document of 2006 noted studies showing effects on kidney functioning at blood lead levels at low as 2–3 μg/dl.

58. Warren, *Brush with Death*, 236.

59. Warren, *Brush with Death*, 221–22.

60. Warren, *Brush with Death*, 239.

61. Lead is still used in bird shot, although this use has come under attack for its environmental impact and is on the wane. The lead shot can be ingested by animals that consume killed birds left in the wild; this can induce lead poisoning in those animals.

62. USGS, "Lead Statistics and Information."

63. Sara M. Bradford, "Lead Smelting Issues in China Could Impact Lead Acid Battery Market," *Frost & Sullivan Market Insight*, October 23, 2009, accessed April 21, 2011, http://www.frost.com/prod/servlet/market-insight-top.pag?docid=183154997.

64. Denworth, *Toxic Truth*, xv.

Chapter 3: Lead Creek

1. The chapter title derives from the principal nickname that was historically used for the South Fork of the Coeur d'Alene River by locals. The term "Lead Creek" presumably owes its origins both to the opaque gray quality of the water body and to the lead tailings that gave the river its distinctive color and texture.

2. Keith R. Long, *Production and Disposal of Mill Tailings in the Coeur d'Alene Mining District, Shoshone County, Idaho: Preliminary Estimates*, Open-File Report 98–595, special report prepared at the request of the United States Geological Survey, 1998, 2.

3. Long, *Production and Disposal of Mill Tailings*, 2. Keith R. Long, John H. DeYoung Jr., and Stephen D. Ludington, *Database of Significant Deposits of Gold, Silver, Copper,*

Lead and Zinc in the United States, Open File Report 98-206A, special report prepared at the request of the United States Geological Survey, 1998.

4. National Research Council, Committee on Superfund Site Assessment and Remediation in the Coeur d'Alene River Basin, *Superfund and Mining Megasites: Lessons from the Coeur d'Alene River Basin* (Washington, DC: National Academies Press, 2005), 43.

5. There are a number of books that use rivers as a lens on the history of an area—Biloine Whiting Young's *River of Conflict, River of Dreams: Three Hundred Years on the Upper Mississippi* (Saint Paul, MN: Pogo, 2004) performs this work with the Upper Mississippi and Bob Deans's *The River Where America Began: A Journey along the James* (Lanham, MD: Rowman and Littlefield, 2007) does it for Virginia's James River—but this chapter does not, except perhaps in a narrow sense. There are also important works of environmental history that place rivers at their center; these include such seminal works as Donald Worster's *Rivers of Empire: Water, Aridity, and the Growth of the American West* (New York: Oxford University Press, 1992) and Richard White's *The Organic Machine: The Remaking of the Columbia River* (New York: Hill and Wang, 1995), and more recent books including Marc Cioc's *The Rhine: An Eco-Biography, 1815–2000* (Seattle: University of Washington Press, 2002) and Sara B. Pritchard's *Confluence: The Nature of Technology and the Remaking of the Rhône* (Cambridge, MA: Harvard University Press, 2011). All of these environmental histories center on a dialectical understanding of the changing interaction between human beings and the natural world. So does this chapter, but in its specific attention to the use of Coeur d'Alene River and its tributaries as an industrial sewer by the mining and milling industry, and the ecological effects of that use, it probably most resembles Cioc's analysis of a far larger river, or Wallace Scot McFarlane's article "Defining a Nuisance: Pollution, Science and Environmental Politics on Maine's Androscoggin River," *Environmental History* 17, no. 2 (2012): 307–35.

6. Nicolas Point, *Wilderness Kingdom: Indian Life in the Rocky Mountains, 1840–1847: The Journals and Paintings of Nicolas Point, S.J.*, trans. Joseph P. Donnelly (New York: Holt, Rhinehart and Winston, 1967), 62.

7. Point, *Wilderness Kingdom*, 180–81.

8. M. M. Ellis for Coeur d'Alene River and Lake Commission, *Special Scientific Report, No. 1, Pollution of the Coeur d'Alene River and Adjacent Waters by Mine Wastes* (Washington, DC: Department of the Interior, Bureau of Fisheries, 1940).

9. Ellis, *Pollution of the Coeur d'Alene.*

10. "Old-Timers Tell How River Valley Looked in Old Days," *Coeur d'Alene Press*, January 10, 1930, 1.

11. Ruby El Hult, *Steamboats in the Timber* (Caldwell, ID: Caxton, 1952), 61.

12. Hult, *Steamboats in the Timber,* 61.

13. "Tailings" refers to waste rock that is left over after the sought-after valuable ore—silver, for instance—has been extracted.

14. Long, "Production and Disposal of Mill Tailings," 2.

15. Long, "Production and Disposal of Mill Tailings," 2.

16. The reagents permit the ores to bind to oily bubbles on the surface of the water. The ores then can be relatively easily scraped off of the water.

17. Elmer Doty Deposition, 1905, in *McCarthy et al. v. Bunker Hill and Sullivan Mining and Concentrating Co. et al.,* MG 130, Box 5, Folder 1, Bunker Hill Records.

18. Plaintiffs' Opening Brief, p. 14, *Doty et al. v. Bunker Hill and Sullivan Mining and Concentrating Company Co. et al.,* MG 130, Box 5, Folder 1, Bunker Hill Records.

19. Plaintiffs' Opening Brief, 14.

20. Plaintiffs' Opening Brief, 14.

21. *Hill v. Standard Mining Company,* in *Reports of Cases Argued and Determined in the Supreme Court of Idaho,* vol. 12 (1906), 227.

22. Casner, "Leaded Waters," 29.

23. Casner, "Leaded Waters," 29.

24. Casner, "Leaded Waters," 29.

25. Casner, "Leaded Waters," 33.

26. Casner, "Leaded Waters," 34.

27. Casner, "Leaded Waters," 40.

28. Casner, "Leaded Waters," 40.

29. The MOA was an organization involving a number of the mining district's larger mining companies that was founded in 1889. The organization marshaled corporate resources to battle labor unions, to struggle collectively against smelters and railroads for lower rates, to lobby the Idaho legislature, to mount collective legal defenses against environmental lawsuits, and to provide for common environmental remediation measures.

30. Frederic L. Quivik, "Of Tailings, Superfund Litigation, and Historians as Experts: *U.S. v. Asarco, et al.* (the Bunker Hill Case in Idaho)," *Public Historian* 26, no. 1 (2004): 90.

31. Casner, "Leaded Waters," 46.

32. Casner, "Leaded Waters," 47.

33. Casner, "Leaded Waters," 47.

34. *Bunker Hill & Sullivan Mining & Concentrating Co. v. Polak,* 7 F.2d 583 (9th Cir. 1925), No. 4461, Circuit Court of Appeals, Ninth Circuit, August 24, 1925, accessed June 16, 2016, http://law.justia.com/cases/federal/appellate-courts/F2/7/583/1509910/.

35. Casner, "Leaded Waters," 48.

36. The MOA's indemnification efforts principally involved Kootenai County residents whose land adjoined the Coeur d'Alene River.

37. Robert Higgs, "Coasian Contracts in the Coeur d'Alene Mining District," *Independent Institute*, June 2, 2004, accessed September 23, 2009, http://www.independent.org/publications/working_papers/article.asp?id=1337. Higgs here is citing an indenture from 1913.

38. Long, "Production and Disposal of Mill Tailings," 2; and Fredric L. Quivik, email message to the author, April 11, 2012 (in the message Quivik referenced files in his possession that contain tables supporting Long's published work). Of that 93 percent of total tailings, the district's largest concentrator of ores, Bunker Hill, produced 28 percent, or close to a third.

39. Long, "Production and Disposal of Mill Tailings," 8. In 1962–1963, the Idaho Department of Transportation dredged tailings from the first impoundment area to use as the roadbed for the section of Interstate 90 that ran through Kellogg.

40. "Test Water of Lake at Boise!," *Coeur d'Alene News*, December 16, 1929, 2.

41. John Knox Coe, "Once It Was a Vale of Plenty But Now . . . ," *Coeur d'Alene Press*, December 23, 1929, 1.

42. Coe, "Once It Was a Vale of Plenty," 1.

43. Coe, "Once It Was a Vale of Plenty," 1.

44. Coe, "Once It Was a Vale of Plenty," 1.

45. Coe, "Once It Was a Vale of Plenty," 1.

46. John Knox Coe, "Valley of Desolation Tells Its Own Story," *Coeur d'Alene Press*, December 26, 1929, 1.

47. Coe, "Pollution of the CDA Can Be Stopped," *Coeur d'Alene Press*, February 14, 1930, 1.

48. *Coeur d'Alene Press*, October 31, 1930, 8.

49. *Coeur d'Alene Press*, October 31, 1930, 8.

50. Until the early 1970s this was common practice for all the communities sitting astride the South Fork and the main stem of the Coeur d'Alene River.

51. Coeur d'Alene River and Lake Commission, *Report and Recommendations*,

52. On a two-to-one vote, sans the signature of Shoshone County Commissioner Taylor, who disagreed with the report's proposed recommendations for the mining industry.

53. Coeur d'Alene River and Lake Commission, *Report and Recommendations of the Coeur d'Alene River and Lake Commission to the Twenty-Second Session of the State Legislature of Idaho* (Boise, 1933), 7.

54. Coeur d'Alene River and Lake Commission, *Report and Recommendations*, 21.

55. Coeur d'Alene River and Lake Commission, *Report and Recommendations,* 21.

56. Coeur d'Alene River and Lake Commission, *Report and Recommendations,* 23.

57. Coeur d'Alene River and Lake Commission, *Report and Recommendations,* 23.

58. Coeur d'Alene River and Lake Commission, *Report and Recommendations,* 48.

59. In the days long before the EPA and Homeland Security, the Department of the Treasury was something of a catchall for a wide variety of federal roles, housing functions from drinking water standards to the Secret Service.

60. Coeur d'Alene River and Lake Commission, *Report and Recommendations,* 48.

61. Coeur d'Alene River and Lake Commission, *Report and Recommendations,* 50.

62. Coeur d'Alene River and Lake Commission, *Report and Recommendations,* 50.

63. Coeur d'Alene River and Lake Commission, *Report and Recommendations,* 50.

64. George Kemmerer, J. F. Bovard, and W. R. Boorman, *Northwestern Lakes of the United States: Biological and Chemical Studies with Reference to Possibilities of Production of Fish,* Bulletin of the Bureau of Fisheries 34 (Washington, DC: Government Printing Office, 1923), 80–81.

65. Kemmerer, Bovard, and Boorman, *Northwestern Lakes of the United States,* 80–81.

66. Kemmerer, Bovard, and Boorman, *Northwestern Lakes of the United States,* 83.

67. Kemmerer, Bovard, and Boorman, *Northwestern Lakes of the United States,* 83.

68. Kemmerer, Bovard, and Boorman, *Northwestern Lakes of the United States,* 51.

69. Ellis, *Pollution of the Coeur d'Alene,* 8.

70. Coeur d'Alene River and Lake Commission, *Report and Recommendations,* 51.

71. Coeur d'Alene River and Lake Commission, *Report and Recommendations,* 52.

72. Coeur d'Alene River and Lake Commission, *Report and Recommendations,* 53.

73. Coeur d'Alene River and Lake Commission, *Report and Recommendations,* 53.

74. Coeur d'Alene River and Lake Commission, *Report and Recommendations,* 53.

75. Ellis, *Pollution of the Coeur d'Alene,* 32.

76. Ellis, *Pollution of the Coeur d'Alene,* 32.

77. Ellis, *Pollution of the Coeur d'Alene,* 34–35.

78. Ellis, *Pollution of the Coeur d'Alene,* 34.

79. Ellis, *Pollution of the Coeur d'Alene,* 44.

80. Ellis, *Pollution of the Coeur d'Alene,* 47.

81. Ellis, *Pollution of the Coeur d'Alene,* 47.

82. That is to say lead ore, or what was released directly from the mills before exposure to weathering action.

83. Ellis, *Pollution of the Coeur d'Alene,* 49.

84. Ellis, *Pollution of the Coeur d'Alene,* 50.

85. Ellis, *Pollution of the Coeur d'Alene,* 51.

86. Ellis, *Pollution of the Coeur d'Alene*, 51.

87. Coeur d'Alene River and Lake Commission, *Report and Recommendations*, 50.

88. Ellis, *Pollution of the Coeur d'Alene*, 54.

89. Ellis, *Pollution of the Coeur d'Alene*, 55.

90. Ellis, *Pollution of the Coeur d'Alene*, 55.

91. Ellis, *Pollution of the Coeur d'Alene*, 25.

92. Keith R. Long reports that between 1932 and 1946, the suction dredge at Cataldo removed more than fifteen million metric tons of tailings-contaminated sediments from the river. See Long, *Production and Disposal of Mill Tailings*, 2.

93. Federal Writers' Project (Idaho), *Idaho: A Guide in Word and Picture* (Caldwell, ID: Caxton, 1937), 172–73.

94. Federal Writers' Project (Idaho), *Idaho*, 332.

95. Federal Writers' Project (Idaho), *Idaho*, 333.

96. Federal Writers' Project (Idaho), *Idaho*, 334.

97. Leonard Dworsky, *Conservation in the United States: A Documentary History: Pollution* (New York: Chelsea House, 1971), 222.

98. The commission pointed out that only 8.5 percent of the sewered population of the Spokane sub-basin had adequate sewage treatment facilities. In the Coeur d'Alenes, where the Coeur d'Alene River and its tributaries had always been used as "sewage sinks," that figure dropped to zero percent.

99. US Public Health Service, Division of Water Pollution Control, *Pacific Northwest Drainage Basins* (Washington, DC: Government Printing Office, 1951), 8.

100. US Public Health Service, *Pacific Northwest Drainage Basins*, 9.

101. US Public Health Service, *Pacific Northwest Drainage Basins*, 26.

102. US Public Health Service, *Pacific Northwest Drainage Basins*, "Summary," x. The year 1951, of course, represented a time of particularly heightened Cold War tensions.

103. US Public Health Service, *Pacific Northwest Drainage Basins*, 11.

104. Keith R. Long, "Tailings under the Bridge: Causes and Consequences of River Disposal of Tailings, Coeur d'Alene Mining Region, 1886 to 1968," *Mining History Journal* 8 (2001): 94.

105. Casner, "Leaded Waters," 136–37.

106. Long, "Tailings under the Bridge," 94.

107. Long, "Tailings under the Bridge," 94; and letter to the editor, *Spokesman-Review*, August 20, 1974, 4. Somewhat ironically, at this time the Pollution Committee of the Coeur d'Alene Lake Property Owners Association actually petitioned the Idaho Department of Public Health to ask the mining companies to continue to discharge tailings into the river, where they would destroy bacteria in the sewage and

thus, it was hoped, impede the prevalence of algae in the lake. The department denied the request.

108. Long, "Tailings under the Bridge," 94.

109. Long, "Tailings under the Bridge," 94.

110. It was deemed cost-prohibitive to link sewer lines to the relative handful of homes in the extremely narrow Burke Canyon. Until late 2007, when septic systems were installed, residents piped their sewage directly into Canyon Creek, a tributary of the South Fork of the Coeur d'Alene.

111. Keith Long estimates that between 1949 and 1997, 16.3 million metric tons of tailings were used as sandfill. See Long, *Production and Disposal of Mill Tailings*, 11.

112. Long, "Tailings under the Bridge," 94.

113. Fred W. Rabe and David C. Flaherty, *The River of Green and Gold: A Pristine Wilderness Dramatically Affected by Man's Discovery of Gold* (Moscow, ID: Idaho Research Foundation, 1974), 50.

114. Rabe and Flaherty, *River of Green and Gold*, 50.

115. Rabe and Flaherty, *River of Green and Gold*, 50.

116. Rabe and Flaherty, *River of Green and Gold*, 55.

117. Rabe and Flaherty, *River of Green and Gold*, 55.

118. Rabe and Flaherty, *River of Green and Gold*, 55.

119. Rabe and Flaherty, *River of Green and Gold*, 56.

120. Rabe and Flaherty, *River of Green and Gold*, 56.

121. Rabe and Flaherty, *River of Green and Gold*, 52.

122. This research, of course, took place before any beneficial aspects of the new settling systems could have been manifested on the river.

123. Rabe and Flaherty, *River of Green and Gold*, 53.

124. The Post Falls Dam was constructed on the Spokane River about ten miles west of the outlet of Lake Coeur d'Alene, in 1906, by the Washington Water Power Company. It raised the level of the Lower Coeur d'Alene River by approximately twelve feet and created many of the lateral lakes along the lower portion of the river. Many farmers and ranchers in the area blamed the dam for causing heavy-metals-laden waters from Lake Coeur d'Alene and the lower river to back up and overflow their lands. Every year in the fall the dam released water, leading to a drawdown of the lateral lakes, which emptied back into the Coeur d'Alene River. In high floods, such as in December 1933 and January 1974, the waters of the river and the lateral lakes merged and became one massive body, spreading heavy metals in the soils and waters over a wide area.

125. Rabe and Flaherty, *River of Green and Gold*, 64.

126. The river was colored this way by the large amounts of "rock flour" from the mining district suspended in the water.

127. Rabe and Flaherty, *River of Green and Gold*, 59.

128. Rabe and Flaherty, *River of Green and Gold*, 59–60.

129. Rabe and Flaherty, *River of Green and Gold*, 60.

130. Rabe and Flaherty, *River of Green and Gold*, 60.

131. Rabe and Flaherty, *River of Green and Gold*, 66.

132. Rabe and Flaherty, *River of Green and Gold*, 66.

133. Rabe and Flaherty, *River of Green and Gold*, 83.

134. Rabe and Flaherty, *River of Green and Gold*, 83.

135. Rabe and Flaherty, *River of Green and Gold*, 84.

136. Rabe and Flaherty, *River of Green and Gold*, 86.

137. Rabe and Flaherty, *River of Green and Gold*, 86.

138. Rabe and Flaherty, *River of Green and Gold*, 50.

139. These were the first waste discharge permits issued under the Clean Air Act of 1972.

140. Rabe and Flaherty, *River of Green and Gold*, 51.

141. US Bureau of Land Management, "Environmental Assessment: Sunshine Sale," 2009, 3, accessed April 16, 2010, http://www.blm.gov/pgdata/etc/medialib/blm/id/nepa/cda_fo/sunshine_precious.Par.71644.File.dat/Sunshine_Precious_Metals_Sale-Decision_FONSI_508.pdf.

142. National Research Council, *Superfund and Mining Megasites*, 79, 81.

143. And available for catch-and-release fishing only.

144. National Research Council, *Superfund and Mining Megasites*, 89.

145. National Research Council, *Superfund and Mining Megasites*, 89.

146. National Research Council, *Superfund and Mining Megasites*, 90.

147. National Research Council, *Superfund and Mining Megasites*, 85.

148. Molly S. Wood and Michael A. Beckwith, *Coeur d'Alene Lake, Idaho: Insights Gained From Limnological Studies of 1991–92 and 2004–06*, prepared in cooperation with the Coeur d'Alene Tribe, US Geological Survey Scientific Investigations Report 2008–5168 (Reston, VA: US Geological Survey, 2008), accessed April 16, 2010, http://pubs.usgs.gov/sir/2008/5168/.

149. Wood and Beckwith, *Coeur d'Alene Lake, Idaho*.

150. Wood and Beckwith, *Coeur d'Alene Lake, Idaho*.

151. Wood and Beckwith, *Coeur d'Alene Lake, Idaho*.

152. National Research Council, *Superfund and Mining Megasites*, 96.

153. In order, for example, to keep the new threats like green algae in check.

154. National Research Council, *Superfund and Mining Megasites*, 96.

155. National Research Council, *Superfund and Mining Megasites*, 96.

156. Although a number of the corporate heads in the mining district were not trained as mining engineers, both Bradley and Easton of Bunker Hill were proud alumni of the University of California, Berkeley. Bradley was a member of the class of 1885 and Easton of the class of 1895.

157. US Department of the Interior, "Coeur d'Alene Basin Final Interim Basin Restoration Plan," April 2007, 33, accessed April 18, 2010, http://www.fws.gov/pacific/ecoservices/envicon/nrda/pdf/CdA_INT%20RES%20FINAL.pdf.

Chapter 4: Foul Humours

1. Berman, *All That Is Solid*, 300.

2. Beck, *Risk Society*, 24.

3. Other works that deal with the interrelations of smelting, economics, pollution control technology, and ecological damage—and that influenced my approach and understanding despite dealing with copper rather than lead and zinc—are Donald MacMillan's *Smoke Wars: Anaconda Copper, Montana Air Pollution, and the Courts, 1890–1924* (Helena: Montana Historical Society Press, 2000), John D. Wirth's *Smelter Smoke in North America: The Politics of Transborder Pollution* (Lawrence: University Press of Kansas, 2000), and LeCain's *Mass Destruction*.

4. Robert N. Bell, "Coeur d'Alene Items, June, 1917," *Engineering and Mining Journal*, July 7, 1917, 6.

5. Bell, "Coeur d'Alene Items," 6.

6. H. O. Hofman, "Metallurgy of Lead in 1916," *Engineering and Mining Journal*, January 6, 1917, 50.

7. Hofman, "Metallurgy of Lead in 1916," 50.

8. "Bunker Hill & Sullivan Smelting Works at Kellogg, Idaho," *Engineering and Mining Journal*, May 13, 1916, 869.

9. With its custom smelter, the company refined much of the lead and silver ores from the Coeur d'Alenes, and additional ores from inside and outside the United States.

10. "Bunker Hill & Sullivan Smelting Works," 868.

11. Aiken, *Idaho's Bunker Hill*, 69.

12. Aiken, *Idaho's Bunker Hill*, 93.

13. Aiken, *Idaho's Bunker Hill*, 93.

14. Beginning in 1925, the company repeatedly, and always successfully, rebuffed smoke damage lawsuits by area farmer Matt Kaiser.

15. MG 367, Box 51, Folder 960, Bunker Hill Records.

16. "Cost of Smoke Damages," 2 October 1940, MG 367, Box 51, Folder 960, Bunker Hill Records.

17. "Cost of Smoke Damages," 2 October 1940.

18. MG 367, Box 51, Folder 960, Bunker Hill Records.

19. Aiken, *Idaho's Bunker Hill*, 89.

20. Aiken, *Idaho's Bunker Hill*, 90.

21. Aiken, *Idaho's Bunker Hill*, 91.

22. While for the sake of convenience I refer above to the plant as "Bunker Hill's," it was in fact legally housed under Bunker's Sullivan Mining Company wing, which was half owned by the Hecla Mining Company.

23. "Zinc Production Started from Sullivan Electrolytic Plant," *Engineering and Mining Journal*, October 20, 1928, 630.

24. The lead smelter produced sulfur dioxide as well, to be sure. The relatively weak strain of sulfur dioxide that came from the smelter, however, was difficult to turn into sulfuric acid. Not until the late 1960s did the company invest in the expensive technology needed to accomplish this task.

25. MacMillan, *Smoke Wars*, 150.

26. MacMillan, *Smoke Wars*, 151.

27. MacMillan, *Smoke Wars*, 151. The arsenic, however, was hardly blameless. It injured and killed many Anaconda-area horses and cows.

28. Wirth, *Smelter Smoke in North America*, 28–29.

29. "Memorandum: Exchange," March 16, 17, 1928, record group (hereafter RG) 95, Box 4, Folder 1928–1929, 1/3, National Archives and Records Administration (hereafter NARA)—Pacific Alaska Region (Seattle).

30. This fear was legitimized by the lawsuit launched by President William Howard Taft's Department of Justice against the Anaconda Company in 1910 over alleged damage caused to the Deer Lodge National Forest by the company's smelter smoke.

31. At this point, that meant the Federal Mining and Smelting Company, Hecla, the Sunshine Mining Company, and the Day Mining Company.

32. General Managers of Bunker Hill, Federal Mining and Smelting Company, and the Sunshine Mining Company to P. D. Hanson, District Forester, District One, Missoula, Montana, 16 August 1946, RG 95, Box 66, Folder "S-Sales, Cd'A—General, 1957 and 1958 and Older Material [2 of 2]," NARA—Seattle.

33. General Managers of Bunker Hill, Federal Mining and Smelting Company and the Sunshine Mining Company to P.D. Hanson, 16 August 1946.

34. The Forest Service divided, and still divides, its management responsibilities into regions. The Coeur d'Alene National Forest was part of the USFS's Region One, headquartered in Missoula, Montana. Regional Forester P. D. Hanson managed all of the forests in Region One, which covered all of Montana, Idaho, Washington, and Oregon, and reported to his superiors at USFS in Washington, DC. The Coeur d'Alene

National Forest was managed on a day-to-day basis by Forest Supervisor K. A. Klehm, headquartered in Coeur d'Alene, Idaho.

35. K. A. Klehm to P. D. Hanson, 1 November 1946, RG 95, Box 66, Folder "S-Sales, Cd'A—General, 1957 and 1958 and Older Material [2 of 2]," NARA—Seattle.

36. It is unclear whether or not Hanson ever sent a final draft of this letter to Robson.

37. P. D. Hanson to Ira A. Robson, December 1946 (?), rough draft, letter, RG 95, Box 66, Folder "S-Sales, Cd'A—General, 1957 and 1958 and Older Material [2 of 2]," NARA—Seattle.

38. Hanson to Robson, December 1946 (?).

39. Hanson to Robson, December 1946 (?).

40. Hanson to Robson, December 1946 (?).

41. Hanson to Robson, December 1946 (?).

42. Charles R. Stark Jr., "Stanly Easton: 64 Years in Mining," unpublished manuscript, 1956, p. 25, MG 5, Box 1, Folder 41, Bunker Hill Records.

43. Stark, "Stanly Easton: 64 Years in Mining," 25.

44. Aiken, *Idaho's Bunker Hill*, 135.

45. "Transcript of Interview with Stanly Easton," October 21, 1957, MG 5, Box 1, Folder 41, Bunker Hill Records.

46. Aiken, *Idaho's Bunker Hill*, 135.

47. Aiken, *Idaho's Bunker Hill*, 144; "Bunker Hill Mining and Smelting Company Builds for the Future," *Spokane Chronicle*, June 24, 1953, 24.

48. US Geological Survey (USGS), "Lead Statistics," accessed October 14, 2010, http://minerals.usgs.gov/ds/2005/140/lead.pdf.

49. Aiken, *Idaho's Bunker Hill*, 135.

50. USGS, "Lead Statistics." Primary production slightly outpaced secondary lead production, which totaled 401,000 (t) in 1951.

51. "New Facilities at Bunker Hill Smelter Step Up Ore Preparations," *Spokesman-Review*, February 18, 1954, 24.

52. Aiken, *Idaho's Bunker Hill*, 135.

53. "This Week in Mining," *Spokesman-Review*, January 11, 1969, 28. Philip J. Landrigan, Edward L. Baker Jr., John L. Earl, and Tscihwa J. Chow, "Biochemical Studies of Lead Isotopes Near a Smelter," lecture, Oak Brook, IL, October 1975, in *Proceedings of the Second International Conference on Stable Isotopes, October 20–23, Oak Brook, Illinois*, 594–620; Art Johnson, "Bunker Hill Said Saved in '70," *Spokesman-Review*, December 2, 1981, 20.

54. Johnson, "Bunker Hill Said Saved in '70."

55. The relatively weak Clean Air Act of 1963 and the amended 1967 version were significantly enhanced and strengthened with the passage of the Clean Air Act of 1970.

56. "Bunker Hill Plans $8 Million Expansion," *Wall Street Journal*, September 16, 1965, 4.

57. Joel Ream, "Bunker Hill Co. Shows Off Costly Clean Air Equipment," *Spokesman-Review*, September 5, 1970, 6.

58. Spokane is approximately eighty-five miles from the Coeur d'Alenes. The ties between the city and the mining district historically have been numerous and strong. D. C. Corbin, builder of the initial narrow-gauge railway into the Coeur d'Alenes, lived in Spokane. A number of the city's other first millionaires, builders of its most opulent nineteenth-century homes, could trace their wealth to the district.

59. Ream, "Bunker Hill Co. Shows Off."

60. Ream, "Bunker Hill Co. Shows Off."

61. "Pollution Control Talk," 1 May 1970, p. 1, MG 413, Box 242, Folder 4547, Bunker Hill Records.

62. "Pollution Control Talk," 2.

63. "Pollution Control Talk," 2.

64. "Pollution Control Talk," 2.

65. "Pollution Control Talk," 2.

66. "Pollution Control Talk," 2. Woodruff makes the same point in Ream, "Bunker Hill Shows Off."

67. "Pollution Control Talk," 2.

68. "Pollution Control Talk," 1.

69. "Pollution Control Talk," 4, 5, 7.

70. "Public Hearing Relative to Application for Extension of Timetables for Control of Air Pollution Submitted by Bunker Hill Company," 3 March 1971, MG 413, Box 242, Folder 4547, Bunker Hill Records.

71. "Public Hearing," 3 March 1971.

72. "Public Hearing," 3 March 1971, 19.

73. "Public Hearing," 3 March 1971, 19.

74. "Public Hearing," 3 March 1971, 30.

75. "Public Hearing," 3 March 1971, 20–21.

76. "Public Hearing," 3 March 1971, 22.

77. "Public Hearing," 3 March 1971, 22. Addressing the meaning of "catastrophe" for the risk society, Ulrich Beck says that "behind all the objectifications, sooner or later the question *of acceptance* arises and with it anew the old question: *how do we wish to live*? . . . The spreading talk of 'catastrophe' is in this sense an objectivized, pointed, radicalized expression that this development is *not wanted*." See Beck, *Risk Society*, 28.

78. "Public Hearing," 3 March 1971, 22.

79. "Public Hearing," 3 March 1971, 23.

80. "Public Hearing," 3 March 1971, 26.

81. Beck, *Risk Society*, 13.

82. "Order and Issuance of Compliance Schedule Environmental Services, Exhibit D," 29 June 1973, p. 1, MG 413, Box 250, Folder 4688, Bunker Hill Records. "Order and Issuance of Compliance Schedule Environmental Services, Exhibit D." The "sulfur recovery rate" refers to the percentage of the sulfur initially entering the zinc and lead plants that the state expected the company to capture or otherwise control.

83. "Order and Issuance of Compliance Schedule Environmental Services," 29 June 1973.

84. "Order and Issuance of Compliance Schedule Environmental Services, Exhibit C," 29 June 1973, p. 1, MG 413, Box 250, Folder 4688, Bunker Hill Records.

85. "Current Situation Bunker Hill vs. EPA Air Pollution," Frank Woodruff to Robert Allen, 18 January 1973, MG 413, Box 242, Folder 4545, Bunker Hill Records.

86. The baghouse fire and the Kellogg lead health epidemic that followed it gained great notoriety for Gulf-Bunker. This incident receives more attention in both the introduction and the next chapter.

87. Bax to Woodruff, 17 September 1973, MG 413, Box 242, Folder 4545, Bunker Hill Records.

88. Woodruff to Bax, 1 October 1973, MG 413, Box 242, Folder 4545, Bunker Hill Records.

89. The hospitalization of these two young Kellogg children in April 1974 set in motion the detection of the community's lead health epidemic that had commenced the previous September with the baghouse fire. This crucial episode in the environmental and social history of the Kellogg area will be dealt with at length in the next chapter.

90. Bax to Gene Baker, 22 May 1974, pp. 1–2, MG 413, Box 243, Folder 4455, Bunker Hill Records. State and federal standards existed for "particulates," as well as sulfur dioxide. *Particulates* essentially refers to gaseous matter released from the Bunker Hill smelter and zinc plant that did not fit into other regulatory categories. The company struggled to comply with state and federal particulate standards for much of the 1970s.

91. Bax to Baker, 22 May 1974, 4.

92. Bax to Baker, 22 May 1974, 4. Baker to Bax, 17 June 1974, MG 413, Box 242, Folder 4545, Bunker Hill Records. The "primary standards" were designed to address human health concerns. Generally more modest "secondary standards" were aimed at protecting plant and animal health.

93. The company's primary sources of sulfur dioxide were the sintering machine at the lead smelter and the flash roaster at the zinc plant.

94. "The Bunker Hill Company Environmental Fact Sheet," December 1976, MG 413, Box 245, Folder 4598, Bunker Hill Records. Some of the sulfur dioxide monitors were run by the State of Idaho.

95. James C. Scott, *Seeing Like a State: How Certain Schemes to Improve the Human Condition Have Failed* (New Haven, CT: Yale University Press, 1998), 80, 78.

96. Scott, *Seeing Like a State*, 78.

97. "Environment III," February 1978 (?), p. 8, MG 413, Box 242, Folder 4602, Bunker Hill Records. If the company were producing greater levels of lead and sinter, it was permitted to expel less sulfur dioxide into the atmosphere, and vice versa. At this time, the federal sulfur dioxide standard was 0.03 ppm.

98. "Bunker Hill Company Environmental Fact Sheet."

99. "Environment III," 8.

100. "Bunker Hill Company Environmental Fact Sheet."

101. "Company Hopes to Keep Smelter in Operation," *Bend (OR) Bulletin*, June 2, 1976, 19.

102. "Company Hopes to Keep Smelter in Operation."

103. "Environment III."

104. "Environment III," 8.

105. "Environment III," 9.

106. Sue Manning, "Bunker Hill—Government Continue Battle," *Spokane Daily Chronicle*, June 8, 1979, 3.

107. Dick Gentry, "Bunker Hill, EPA Settle," *Spokane Daily Chronicle*, June 8, 1979, 4.

108. The Kellogg area, of course, only *looked* like a desert. It had come to bear such a resemblance following decades of sulfur dioxide and heavy metals pollution. As one moved farther away from the mining district, the landscape began again to resemble the heavily forested, verdant scenery typical of the moist Pacific Northwest region to which the Coeur d'Alenes belong.

109. "Bunker Hill Company Environmental Fact Sheet."

110. "Bunker Hill Company Environmental Fact Sheet."

111. "Environment III."

112. "Bunker Hill Company Environmental Fact Sheet."

113. "Bunker Hill Company Environmental Fact Sheet."

114. "Revegetation Job Praised," *Spokane Daily Chronicle*, July 20, 1977, 40.

115. "Revegetation under Way in Idaho," *Toledo Blade*, November 27, 1975, C20.

116. J. H. Halley to Frank Woodruff, interoffice memo, 16 May 1975, MG 413, Box 245, Folder 4591, Bunker Hill Records. For a cost, the company also made hydroseeding equipment available to residents of those communities who wished to put down grass seed.

117. For example, in the first half of 1975, 15,000 willow shoots and 1,500 shrubs were planted, and the spring planting of 1977 included 10,500 shrubs. By 1977, 40,000 willows had been planted.

118. Halley to Woodruff, interoffice memo, 18 July 1975, MG 413, Box 245, Folder 4591, Bunker Hill Records.

119. "Environment III," 4.

120. Julie Titone, "Corps Renews Battle for Bunker Hill: Choppers Bombard Superfund Site in Effort to Revegetate Barren Slopes," *Spokesman Review*, September 25, 1998, 1.

121. Personal visit to Kellogg, March 2010.

Chapter 5: Raining Down Poison

Epigraphs: Beck, *Risk Society*, 23; Berman, *All That Is Solid*, 100.

1. Julie Titone, "Monumental Cleanup: Smelterville Baghouse Was Source of Lead Contamination That Poisoned Silver Valley," *Spokesman-Review*, March 21, 1995, B4.

2. It was clear from the time of the fire that it would take months to repair the baghouse. This was due primarily to a major supply shortage of replacement bags.

3. It probably never will be known exactly which company officials made the decision to keep the smelter running following the baghouse fire. By the summer of 1973, Bunker Hill had been a wholly owned subsidiary of Houston-based Gulf Resources and Chemical Corporation for five years. Top Gulf officials such as CEO Robert Allen and vice-president Frank Woodruff always maintained that they had nothing to do with the decision, that it was a routine managerial decision made at the plant level in Kellogg.

4. Aiken, *Idaho's Bunker Hill*, 180.

5. "Lead Concentration Worst in Nation," *Spokesman-Review*, September 23, 1981, 3.

6. David Bond, "Smelter Firm Put Price Tag on Lead Risk," *Spokesman-Review*, July 21, 1990, A1.

7. "Lead Concentration Worst in Nation."

8. "Smelter," A6. It remains unclear who in the Gulf-Bunker hierarchy authorized this decision, although in this case as well Gulf executives denied having had anything to do with it.

9. "Smelter," A6. The literature on human health and the environment is vast and rich. The cornerstone of this canon is undoubtedly Rachel Carson's *Silent Spring* (1962), which limned the risk posed to the web of life by pesticides. The fundamental insight of this body of literature is that human bodies are inseparable from the environments they inhabit, that environmental concerns and human health are inescapably in-

terrelated. Another primary insight is that while humans undoubtedly affect environments, these altered environments in turn shape humans. In addition to *Silent Spring*, the literature that has particularly influenced the writing of this chapter includes Linda Nash's *Inescapable Ecologies: A History of Environment, Disease, and Knowledge* (2006), which examines the 160-year relationship between people and local environments in California's Central Valley, culminating with the study of the varying perceptions of a significant cancer cluster in pesticide-soaked McFarland; Gregg Mitman's *Landscapes of Exposure: Knowledge and Illness in Modern Environments* (Chicago: University of Chicago Press, 2004), which covers similar terrain; the chapter "The Toxic Environment" in Samuel P. Hays's *Beauty, Health, and Permanence: Environmental Politics in the United States, 1955–1985* (Cambridge, UK: Cambridge University Press, 1987), which provides a strong overview of controversies over human-made environmental toxins; Andrew Hurley's fascinating study of the politics of pollution in Gary, Indiana, *Environmental Inequalities: Class, Race, and Industrial Pollution in Gary, Indiana, 1945–1980* (Chapel Hill: University of North Carolina Press, 1995); Andrea Peacock's poignant story of the asbestos poisoning of a small Montana town by W. R. Grace Corporation, *Libby, Montana: Asbestos and the Deadly Silence of an American Corporation* (Boulder, CO: Johnson, 2003); Ishimure Michiko's *Paradise in the Sea of Sorrow: Our Minamata Disease* (Ann Arbor: Center for Japanese Studies at the University of Michigan, 1990), the story of the mercury poisoning of a Japanese community; Brett L. Walker's powerful indictment of the toll that Japan's intensive bouts of industrial modernization have exacted on the bodies of its people, *Toxic Archipelago: A History of Industrial Disease in Japan* (Seattle: University of Washington Press, 2010); and Gerald Markowitz and David Rosner's *Deceit and Denial: The Deadly Politics of Industrial Pollution* (Berkeley: University of California Press, 2002), a muckraking-style history that details outstanding examples of corporate harm to human health.

10. Beck, *Risk Society*, 35.

11. Berman, *All That Is Solid*, 74.

12. Aiken, *Idaho's Bunker Hill*, 172.

13. "Memorandum of Authorities in Support of Plaintiffs' Motion for Partial Summary Judgment," *Edna Grace Yoss, et al., v. The Bunker Hill Company, et al.*, 13 April 1981, p. 21, MG 413, Box 231, Folder 4434, Bunker Hill Records.

14. Although not considered as accurate as blood lead screenings, urine tests remained fairly common in the early 1970s.

15. "Memorandum of Authorities in Support of Plaintiffs' Motion," 21.

16. "Memorandum of Authorities in Support of Plaintiffs' Motion," 21.

17. "Smelter."

18. "Smelter."

19. "Smelter."

20. Aiken, *Idaho's Bunker Hill*, 182.

21. Ian von Lindern and George Dekan to Jerry Cobb, 6 May 1974, MG 413, Box 236, Folder 4424, Bunker Hill Records.

22. Von Lindern and Dekan to Cobb, 6 May 1974.

23. Statement by Shoshone Lead Project Technical Committee, 4 September 1975, MG 413, Box 236, Folder 4415, Bunker Hill Records.

24. Statement by Shoshone Lead Project Technical Committee.

25. Dr. Philip J. Landrigan, testimony before the Subcommittee on Health and the Environment (date and chamber of Congress unknown), MG 413, Box 236, Folder 4435, Bunker Hill Records. The five study areas were within radii of, respectively, 1 mile, 2.5 miles, 6 miles, 15 miles, and 20 miles from the smelter.

26. An "elevated" blood lead level means that it is considered medically unsafe and of cause for concern. In 1974, a blood lead level of 40 µg/dl was the federal standard for unsafe levels of lead in the bloodstream. Since 1991, blood lead levels have been considered to be elevated at one-quarter the former standard, or 10 µg/dl.

27. Landrigan, testimony before the Subcommittee on Health and the Environment.

28. "Memorandum of Authorities in Support of Plaintiffs' Motion," 11.

29. "Memorandum of Authorities in Support of Plaintiffs' Motion," 11.

30. Statement by Shoshone Lead Project Technical Committee.

31. Von Lindern to Stokes, 29 April 1975, MG 413, Box 241, Folder 4530, Bunker Hill Records. By comparison, the 1978 federal standards for ambient air lead set concentrations of 1.5 $\mu g/m^3$ as the maximum safe level, less than one-sixth of the level measured around Kellogg in late 1973 and early 1974.

32. Baker to Jim Keane, interoffice memo, 8 June 1977, MG 413, Box 236, Folder 4415, Bunker Hill Records. In March 1974, the month of highest emissions, 96.1 tons of lead went up the smelter stack.

33. Statement by Shoshone Lead Project Technical Committee.

34. Baker to Dale Simpson, 17 July 1974, MG 413, Box 234, Folder 4400, Bunker Hill Records.

35. Baker to Dale Simpson, 17 July 1974; 26 February 1975, MG 413, Box 234, Folder 4400, Bunker Hill Records.

36. T. D. Tennent to Lawrence Snyder, 11 July 1975, MG 413, Box 234, Folder 4400, Bunker Hill Records. In this missive Tennent also specifically admonished Snyder to get his other horse off the land where one of his horses had just perished.

37. Statement by Shoshone Lead Project Technical Committee.

38. Glen Wegner, ed., *Shoshone Lead Health Project: Work Summary* (Boise: Idaho Department of Health and Welfare, 1976), 3.

39. Wegner, *Shoshone Lead Health Project*, 3.

40. Wegner, *Shoshone Lead Health Project*, 5.

41. Wegner, *Shoshone Lead Health Project*, 8.

42. Wegner, *Shoshone Lead Health Project*, 12.

43. Wegner, *Shoshone Lead Health Project*, 34. Mean blood leads also were found to follow a logarithmic pattern based upon amount of an area group's lead exposure; this was in keeping with the previously discovered pattern of strontium-90 radiation poisoning cases.

44. Wegner, *Shoshone Lead Health Project*, 74.

45. Wegner, *Shoshone Lead Health Project*, 183. Yankel and von Lindern noted the propinquity of the fact that evidently homegrown vegetables did not constitute a major portion of the Silver Valley diet and recommended that no one eat the vegetables grown in the valley. The mean blood lead for children in the area within one mile of the smelter stack was 68 µg/100 ml. For the area between a mile and two and a half miles from the stack, the mean blood lead was 49.1 µg/100 ml. At the Silver King Elementary School, located 0.3 miles from the stack, monthly air lead averages reached higher than 20 $\mu g/m^3$ (more than ten times the 1978 EPA standard of 1.5 $\mu g/m^3$). Soil lead averages at Silver King reached over 10,000 ppm in July 1974.

46. Wegner, *Shoshone Lead Health Project*, 73.

47. Wegner, *Shoshone Lead Health Project*, 73.

48. Wegner, *Shoshone Lead Health Project*, 76.

49. Wegner, *Shoshone Lead Health Project*, 9. Six high-volume air samplers were installed by the Idaho Department of Health and Welfare in April–May 1974 when increased concern prompted by the Thomas girls' illness prompted the department's efforts to make the Silver Valley environment far more legible; the other three dated to 1971. Of the first three samplers two were located in Kellogg and one in Smelterville. The latter six were spread out all over the valley, ranging from Mullan (18.5 miles from the stack) to the Silver King Elementary School (0.3 miles distant).

50. Wegner, *Shoshone Lead Health Project*, 76.

51. Wegner, *Shoshone Lead Health Project*, 76.

52. Wegner, *Shoshone Lead Health Project*, 76. Turning to other heavy metals found in high concentrations throughout the Silver Valley—cadmium, zinc, arsenic, antimony, mercury—the scientists concluded that cadmium, a powerful known toxin, likely posed the greatest danger to human health. Like lead, cadmium and the other heavy metals were found in decreasing concentrations with greater distance from the smelter.

53. Wegner, *Shoshone Lead Health Project*, 82.

54. Wegner, *Shoshone Lead Health Project*, 78–79.

55. Wegner, *Shoshone Lead Health Project,* 102.

56. Wegner, *Shoshone Lead Health Project,* 103. Lead oxides break down much more easily in the environment than do lead sulfides and thus are more readily bioavailable, and more toxic, to humans and other forms of life.

57. Wegner, *Shoshone Lead Health Project,* 96. The five areas corresponded to radii in progressively greater distances from the smelter: Area 1 lay within a one-mile radius of the smelter while Area 5, centered at Mullan, lay fifteen to twenty miles distant. Race was not used as a determinant, as the Silver Valley was (and remains) almost uniformly white.

58. Wegner, *Shoshone Lead Health Project,* 97. The mean blood lead for the 174 children of smelter workers living in Areas 1–4 was 55.1 µg/dl, while that for the 233 children of underground miners living in the same areas was 45.8 µg/dl. The mean blood lead for all other children living in those areas was 42.1 µg/dl.

59. Wegner, *Shoshone Lead Health Project,* 97.

60. Like von Lindern and Yankel, Landrigan's research led him to conclude that the consumption of household paint did not constitute a significant means by which Shoshone County children had become leaded.

61. *Nerve conduction velocity* refers to the speed at which electrical impulses move through a nerve.

62. Wegner, *Shoshone Lead Health Project,* 100.

63. Wegner, *Shoshone Lead Health Project,* 185. This is von Lindern's and Yankel's characterization.

64. Wegner, *Shoshone Lead Health Project,* 177. Annual tests were recommended for children with blood lead levels of 40–59 µg/dl. For those with blood lead levels of 60 µg/dl and higher, biannual tests were recommended along with home visits by Dr. Panke and Dr. Mather, aimed at advising parents on how to decrease the amount of available lead in the home environment. Areas 1 and 2 essentially corresponded to the communities of Smelterville and Kellogg, respectively.

65. Wegner, *Shoshone Lead Health Project,* 179.

66. Wegner, *Shoshone Lead Health Project,* 180–81.

67. The company also, beginning in 1976, needed to respond to new Office of Safety and Health Administration (OSHA) rules for airborne lead in the workplace. These rules affected the operation of Bunker Hill's lead smelter.

68. This of course was subject to approval by EPA. The agency also watched over and backstopped the states' monitoring programs.

69. A micron is one-millionth of a meter. A human hair, by comparison, is one hundred microns wide. Particulates smaller than 1.5 microns are of particular concern because they readily can be absorbed into lung tissue.

70. Keith Dahlberg to Idaho Pollution Commission, 25 September 1971, MG 413, Box 243, Folder 4551, Bunker Hill Records.

71. "Idaho Department of Environmental Protection and Health, Air Pollution Abatement Program Time Schedule for Particulate Control: The Bunker Hill Company, February 6, 1973," MG 413, Box 243, Folder 4552, Bunker Hill Records.

72. "Idaho Department of Environmental Protection and Health, Air Pollution Abatement Program Time Schedule," "Exhibit A," 1.

73. Gene Baker to Lee Stokes, 14 May 1974, MG 413, Box 250, Folder 4688, Bunker Hill Records.

74. Bax to Baker, 22 May 1974, MG 413, Box 250, Folder 4688, Bunker Hill Records.

75. Bax to Baker, 22 May 1974.

76. Bax to Baker, 22 May 1974.

77. Bax to Baker, 22 May 1974.

78. Beck, *Risk Society*, 24.

79. Hooper Report, "Concepts for Control of Lead Emissions from the Bunker Hill Smelter," 5 December 1974, MG 413, Box 243, Folder 4553, Bunker Hill Records.

80. Mark H. Hooper, testimony before the Idaho State Board of Health and Welfare, 18 August 1975, p. 8, MG 413, Box 243, Folder 4557, Bunker Hill Records.

81. Hooper, testimony, 18 August 1975, 8.

82. Hooper, testimony, 18 August 1975, 9.

83. Hooper Report, "Concepts for Control of Lead Emissions."

84. Hooper, testimony, 18 August 1975, 9.

85. Hooper Report, "Concepts for Control of Lead Emissions."

86. Hooper Report, "Concepts for Control of Lead Emissions." Scrubbers constitute a technology that generally employs streams of water to separate out pollutants from desired metals.

87. Hooper, testimony, 18 August 1975, 10.

88. Hooper, testimony, 18 August 1975, 10.

89. "Bunker Hill Company Environmental Fact Sheet," 6.

90. Beck, *Risk Society*, 28.

91. Beck, *Risk Society*, 28.

92. "Environment III," February 1978 (?), p. 12, MG 413, Box 242, Folder 4602, Bunker Hill Records.

93. "Environment III," 12.

94. "Environment III," 12. Maintaining that it would cost $350,000 "to further reduce emissions" from the fuming furnace and that the expenditure would result in only a 1 percent reduction in "lead particulate emissions," the company charged that the EPA mandate was fundamentally unreasonable.

95. "Environment III," 12.

96. "Environment III."

97. PES study of particulate distribution in the Silver Valley, October 1977–September 1978, Table 5-4, MG 413, Box 236, Folder 4437, Bunker Hill Records.

98. DECS Environmental Specialist George Dekan to Bob Olson, memorandum, 29 October 1973, MG 413, Box 243, Folder 4552, Bunker Hill Records.

99. PES study of particulate distribution in the Silver Valley. The environmental firm used DHW's estimates of background particulate concentrations in making its conclusions.

100. PES study of particulate distribution in the Silver Valley, Table 5-6.

101. PES study of particulate distribution in the Silver Valley, Table 5-2. Both of the tall stacks were online by August 1977.

102. PES study of particulate distribution in the Silver Valley, Table 8-4. Geometric means for Silver King and Smelterville were 65 and 72 $\mu g/m^3$, respectively.

103. Sue Manning, "Bunker Hill—Government Continue Battles," *Spokesman Daily Chronicle*, June 8, 1979, 3.

104. "Environment III," 12.

105. Jorge A. Roche Lizarraga, "Air Quality in the Bunker Hill Area," 1980 (?), Table 1, MG 413, Box 242, Folder 4550, Bunker Hill Collection. This data reflected lead-in-air concentrations measured as a geometric average.

106. Lizarraga, "Air Quality in the Bunker Hill Area," Table 1. The "fairly typical" month used was December.

107. Lizarraga, "Air Quality in the Bunker Hill Area," 4. The date of that triple-digit reading was October 28, 1973.

108. PES study of particulate distribution in the Kellogg area, Table 5-2. The study compared lead levels from two periods: July 1974–June 1977 and October 1977–September 1978.

109. PES study of particulate distribution in the Kellogg area, Table 5-1.

110. As previously discussed, this was true to a greater extent in locales farther distant from the smelter.

111. PES study of particulate distribution in the Kellogg area, Table 8-8.

112. PES study of particulate distribution in the Kellogg area, Table 8-8.

113. Among other things, the LIA, whose suit Bunker joined, pointed to the fact that in major cities such as Los Angeles, ambient air lead levels were high above the EPA standard, due primarily to the effects of leaded gasoline. In Los Angeles, for instance, ambient levels averaged nearly double the 1.5 $\mu g/m^3$ rule.

114. Aiken, *Idaho's Bunker Hill*, 198–99.

115. "Bunker Snubs EPA," *Spokesman- Review*, April 8, 1980, A6.

116. "Plan Will Take 8 Months," *Spokane Daily Chronicle*, September 7, 1981, 3.

117. "Plan Will Take 8 Months," 3.

118. "Third Party Complaint," 1 December 1977, p. 23, MG 413, Box 231, Folder 4339, Bunker Hill Records. Recall that any blood lead of 40 μg/dl was, at the time, considered elevated. Anything 80 μg/dl or above was deemed "frankly pathological" and *prima facie* grounds for immediate hospitalization.

119. To respect the privacy of the Dennis children (now middle-aged adults), I am not going to name them individually beyond giving their 1974 blood lead readings.

120. Dr. Robert J. Gregory, "Notes on Yoss and Dennis Children," 22 October 1980, MG 413, Box 236, Folder 4437, Bunker Hill Records.

121. J. H. Lollar, "June Management Meeting Minutes," memorandum, 11 June 1977, p. 3, MG 413, Box 242, Folder 4602, Bunker Hill Records.

122. "Third Party Complaint," 4.

123. Deposition Exhibit Number 7, MG 413, Box 236, Folder 4437, Bunker Hill Records.

124. Deposition Exhibit Number 7.

125. Deposition Exhibit Number 7; Gregory, "Notes on Yoss and Dennis Children."

126. Honora Simpson, "Summary," (of June 22–23, 1980, tests in Chicago), MG 413, Box 244, Folder 4572, Bunker Hill Records.

127. Deposition Exhibit Number 7.

128. Lattimer, "Pediatric Consultant's Report," January 7, 1977, MG 413, Box 244, Folder 4572, Bunker Hill Records.

129. Lattimer, "Pediatric Consultant's Report."

130. Simpson, "Summary."

131. Deposition Exhibit Number 7.

132. While Dr. Gregory, the University of Idaho psychologist, labeled three of these children as having "test results essentially normal" and one as having a "test battery more typical of normal than abnormal psychological function" in his October 1980 summary, Dr. Dodrill, in his 1979 report, diagnosed two of these four as having "mild" or "mild but definite" brain impairments. See Gregory's comments in Gregory, "Notes on Yoss and Dennis Children"; and Dodrill's in Deposition Exhibit Number 7.

133. "% Reduction Earning Based on Janet Mott," Exhibit 10, MG 413, Box 243, Folder 4556, Bunker Hill Records.

134. "Lead Poisoning Suit Settled," *Tri-City Herald*, February 2, 1982, 4.

135. "Lead Poisoning Suit Settled," 4. Intriguingly, the settlement closely approximated the $8 million of dividends paid by Bunker to Gulf Resources in 1974, the year following the baghouse fire.

136. Beck, *Risk Society*, 23.

137. Susan Drumheller and Craig Welsh, "Children Left in the Dust: Lead Affected Generation of Silver Valley Children," *Spokesman-Review*, March 9, 1997, A1.

138. Drumheller and Welsh, "Children Left in the Dust." Lead in the bone is a good measurement of lifetime exposure to lead.

139. Drumheller and Welsh, "Children Left in the Dust."

140. Nicholas K. Geranios, "Victims of North Idaho Pollution Still Suffer Physical, Emotional Ills," *Missoulian*, September 5, 2010.

141. Drumheller and Welsh, "Children Left in the Dust."

142. Drumheller and Welsh, "Children Left in the Dust."

143. Drumheller and Welsh, "Children Left in the Dust."

144. "Interrogatories to Bunker Hill Company," 16 September 1975, p. 10, MG 413, Box 236, Folder 4415, Bunker Hill Records.

145. "Memorandum of Authorities in Support of Plaintiffs' Motion for Summary Judgment," 23, MG 413, Box 231, Folder 4343, Bunker Hill Records.

146. "Memorandum of Law in Opposition to Defendants' Motion for a Protective Order: Mean Blood Leads for Cohort Children by Area & Year," MG 413, Box 231, Folder 4341, Bunker Hill Records.

147. "Memorandum of Law in Opposition to Defendants' Motion."

148. Idaho Department of Health and Welfare, *Kellogg Revisited—1983 Childhood Blood Lead and Environmental Status Report*, p. 1, MG 413, Box 253, Folder 4716, Bunker Hill Records.

149. "Status of Blood Lead Determinations in Shoshone County April 1980, Table I," MG 413, Box 236, Folder 4424, Bunker Hill Records.

150. William H. Foege to Edward S. Gallagher, 8 April 1980, p. 2, MG 413, Box 234, Folder 4402, Bunker Hill Records.

151. "Stenographic Transcript of Hearings before the Subcommittee on Health and Environment of the Committee on Interstate and Foreign Commerce," 27 November 1979, p. 74, MG 413, Box 246, Folder 4617, Bunker Hill Records.

152. "Stenographic Transcript," 105–8.

153. Bill Graves, "Kellogg Lead Controversy Still Raging," *Coeur d'Alene Press*, October 20, 1979. In addition to Landrigan, another Shoshone Project contributor, psychologist Robert Gregory, broke ranks with the project's findings. In October 1980, Gregory, speaking on a television broadcast, said that "even in 1974" the project's conclusion that no Shoshone County child had been clinically impaired by lead exposure had been wrong, that "it's a virtual certainty" that some of the children had suffered minor neurological impairment, and that some of the children might manifest further decline in intellectual ability in the future. Adding medical and scientific credibility to

positions such as those taken by Landrigan and Gregory was the publication in the *New England Journal of Medicine* in 1979 of "Deficits in Psychologic and Classroom Performance of Children with Elevated Dentine Lead Levels" (for more detail, see discussion of the article in chapter 1, "The Useful Metal"). The article, authored by Dr. Herbert L. Needleman, a pathbreaking lead health researcher, detailed the IQ and classroom deficits of a group of Boston-area children, all of whom, like Kellogg-area children, manifested no overt symptoms of lead poisoning. The Boston children, however, had median blood lead levels that were but fractions of those experienced by Kellogg area children in 1974.

154. "Shoshone Lead Grant," 1, MG 413, Box 250, Folder 4670, Bunker Hill Records.

155. "Shoshone Lead Grant," 1.

156. Idaho Department of Health and Welfare, *Kellogg Revisited*, 17.

157. Idaho Department of Health and Welfare, *Kellogg Revisited*, 23.

158. Bunker Hill research in 1972 detected soil lead means of greater than 5,000 ppm and 1974 studies found levels of greater than 10,000 ppm at Silver King school.

159. "Lead Victims Say Their Health Was Hurt," *Seattle Times*, September 5, 2010. In 1991, the CDC again defined "unsafe levels" of blood lead downward, to 10 μg/dl. This remains the definition of "undue lead absorption" in the United States.

160. "The Box" runs roughly from Elizabeth Park in the east to Pinehurst in the west.

161. Suzanne Jacobsen, "Blood Testing Efforts to Heighten," *Shoshone News-Press*, August 31, 2007.

162. "EPA: Lawn Remediation Complete," *Shoshone News-Press*, July 10, 2008.

163. "Lead Victims Say Their Health Was Hurt."

164. Sean Garnire, "IDEQ Seasonal Effort Ends," *Shoshone News-Press*, October 11, 2007. National blood lead levels also had dropped considerably, though not as far or as fast as those of the Silver Valley. The phasing out of leaded gasoline in the 1980s had a good deal to do with this outcome nationally. The average blood lead levels of American fifteen-year-olds declined from 15.0 μg/dl in the 1976–1980 period to 2.7 μg/dl in the 1991–1994 period.

165. William G. Robbins, *Colony and Empire: The Capitalist Transformation of the American West* (Lawrence: University Press of Kansas, 1994), 163.

166. Made famous by Dan Fagin's recent book, *Toms River: A Story of Science and Salvation* (New York: Bantam Books, 2013).

Chapter 6: On Lead's Front Lines

1. There are many works that deal with the subject of mining accidents and health-related issues (e.g., silicosis), but relatively few that focus on the issue of worker health

in smelting facilities. That said, mining long has been, and remains, a highly dangerous industry in the Coeur d'Alenes. Silicosis was not uncommon among miners, nor were mining injuries and deaths. Mining accidents have claimed the lives of approximately 976 miners during the district's 130-year history. The worst, and most notorious, mining disaster occurred in the Sunshine Mine on May 2, 1972, when a fire caused the death of 91 miners by smoke inhalation and carbon monoxide poisoning. To give some idea of the rate of accident and death in the district, in July 1912, Bunker Hill manager Stanly Easton was distressed to report to company president F. W. Bradley that there had been twelve deaths at the Bunker Hill Mine over the previous eighteen months. These twelve deaths represented two-thirds of the total deaths (eighteen) over the previous decade at the mine. During its most dangerous period, 1913–1914, the mining district averaged two deaths per month. Around 1930, the district's death and accident rates began to drop precipitously due to improvements made in mine safety. Looking at two of the district's largest mines, the Bunker Hill Mine suffered 97 accidents and no deaths in 1924. Its rate of accidents per 1,000 shifts was 0.1783. By comparison, the Hecla Mine experienced a rate of 2.51 accidents per 1,000 shifts. During Easton's long tenure at Bunker Hill (1902–1957), the company made mine safety a significant priority, consistently investing in the latest safety technology and boasting a mine rescue safety team that competed nationally. Major improvements in overall safety at the district's mines, however, particularly from 1930 onward, could not negate the reality that mining has taken a high toll on the communities of the Coeur d'Alenes, in the form of killed and incapacitated community and family members and their lost earning power.

2. Arthur F. McEvoy, "Working Environments: An Ecological Approach to Industrial Health and Safety," in "Snapshots of a Discipline: Selected Proceedings from the Conference on Critical Problems and Research Frontiers in the History of Technology, Madison, Wisconsin, October 30–November 3, 1991," supplement, *Technology and Culture* 36, no. 2 (April 1995): S146–47.

3. McEvoy, "Working Environments," S164.

4. In addition to McEvoy's pathbreaking essay, and the texts mentioned as influences in the previous chapter, this chapter has been influenced by *Dying for Work: Workers' Safety and Health in Twentieth-Century America*, a trenchant collection of essays addressing a variety of historical worker health and safety situations, edited by David Rosner and Gerald Markowitz.

5. Easton to Bradley, 20 July 1917, MG 367, Box 10, Folder 129, Bunker Hill Records.

6. Easton to Labarthe, 22 August 1918, MG 367, Box 12, Folder 155, Bunker Hill Records.

7. Sayers to Easton, 16 July 1918, MG 4, Box 12, Folder 155, Bunker Hill Records.

8. Sayers to Easton, 16 July 1918; Sayers to Easton, 15 July 1918, MG 4, Box 12, Folder 155, Bunker Hill Records.

9. Sayers to Easton, 15 July 1918, 2.

10. Sayers to Easton, 15 July 1918, 4–5.

11. Sayers to Easton, 15 July 1918, 4–5.

12. Sayers to Easton, 15 July 1918, 4.

13. Aiken, *Idaho's Bunker Hill*, 93.

14. According to the "assumption of risk" doctrine, if a worker was properly informed of the inherent risks of his occupational workplace and still chose to pursue his occupation in that place, his employer could not be held liable for any sickness or injury that befell him while working there. The "fellow servant rule" placed the burden of maintaining a safe workplace upon one's fellow workers, not upon the employer.

15. Easton to Bradley, 8 October 1924, p. 1, MG 367, Box 12, Folder 149, Bunker Hill Records.

16. Easton to Bradley, 8 October 1924, 1. Before the adoption of a Workmen's Compensation Act by the State of Idaho in 1917, Bunker Hill adopted the same policy toward "worthy employees" who suffered mining injuries and for the widows of those who had been killed in the Bunker Hill Mine.

17. Easton to Bradley, 14 October 1924, p. 1, MG 367, Box 12, Folder 149, Bunker Hill Records.

18. Aiken, *Idaho's Bunker Hill*, 93.

19. Aiken, *Idaho's Bunker Hill*, 95.

20. Rickard, *Bunker Hill Enterprise*, 107.

21. Rickard, *Bunker Hill Enterprise*, 94; "Solarium Proves Success," *Spokesman-Review*, September 26, 1929, 11.

22. Aiken, *Idaho's Bunker Hill*, 94.

23. Aiken, *Idaho's Bunker Hill*, 94; and "See Model Mines," *Lewiston Morning Tribune*, October 5, 1929, 3.

24. Aiken, *Idaho's Bunker Hill*, 11.

25. Aiken, *Idaho's Bunker Hill*, 11.

26. Aiken, *Idaho's Bunker Hill*, 48; and Easton to Bradley, 10 October 1907, MG 367, Box 10, Folder 110, Bunker Hill Records.

27. "The Bunker Hill and Sullivan Enterprise Today," *Engineering and Mining Journal*, August 1939, 82.

28. "Bunker Hill and Sullivan Enterprise," 82.

29. Easton to Bradley, 3 December 1917, p. 2, MG 367, Box 10, Folder 130, Bunker Hill Records. In 1939, an occupational diseases section that covered silicosis was added to the act.

30. Easton to Bradley, 3 December 1917, 2; Easton to Bradley, 29 December 1917, MG 367, Box 10, Folder 130, Bunker Hill Records.

31. Aiken, *Idaho's Bunker Hill*, 93.

32. "Bunker Hill and Sullivan Enterprise," 79. Dust of course was also a problem for miners, as it had been tied to the incidence of silicosis. Under Bunker's new rules, however, efforts also were made to address the problem of mine dust. Water began to be used to wet down walls and muck piles before shoveling, and blasting was postponed until the ends of shifts.

33. Among other wartime uses, zinc was combined with copper to make brass shell casings.

34. Aiken, *Idaho's Bunker Hill*, 122.

35. Initial efforts at unionizing the company, led by the Western Federation of Miners, had ushered large-scale outbreaks of violence and recrimination in the district in 1894 and 1899. Under the fiercely antiunion leadership of Bradley and Easton, Bunker had held independent unionism at bay while sponsoring the Wardner Industrial Union, a "company union," as its surrogate. By the early 1940s, however, the effects of the New Deal's prolabor stance, federal policy changes such as the Wagner Labor Relations Act of 1937, and wartime control of industry proved sufficient to overwhelm the company's long-standing opposition to trade unionism.

36. Aiken, *Idaho's Bunker Hill*, 118.

37. Aiken, *Idaho's Bunker Hill*, 119.

38. Aiken, *Idaho's Bunker Hill*, 121.

39. Aiken, *Idaho's Bunker Hill*, 121.

40. Aiken, *Idaho's Bunker Hill*, 137. Easton stepped down as Bunker Hill president in 1954 and Bradley assumed that position on March 29, 1955. Longtime company executive J. B. Haffner served as president in the interim.

41. Aiken, *Idaho's Bunker Hill*, 142.

42. Aiken, *Idaho's Bunker Hill*, 142–43.

43. Warning, 16 January 1954, MG 367, Box 73, Folder 1525, Bunker Hill Records.

44. Aiken, *Idaho's Bunker Hill*, 143.

45. "Bunker Hill Sets $15 Million Expansion," *Kellogg Evening News*, January 15, 1957, 1.

46. Aiken, *Idaho's Bunker Hill*, 145.

47. Beck, *Risk Society*, 13. Beck contrasts this stage of industrial society with the subsequent stage, which he has dubbed *advanced modernity*, in which concern over the risks of industry come to outweigh the focus upon its benefits.

48. Aiken, *Idaho's Bunker Hill*, 145.

49. Aiken, *Idaho's Bunker Hill*, 150.

50. Aiken, *Idaho's Bunker Hill*, 154.

51. A CIO investigation in 1950 confirmed that some of the IUMMSW's national leadership were members of the Communist Party. Following this investigation, the CIO expelled Mine-Mill from its ranks. With Cold War tensions on the rise in the early 1950s, the issue of Mine-Mill's connections to "international Communism" became more politically potent in the Coeur d'Alenes.

52. Aiken, *Idaho's Bunker Hill*, 138.

53. Aiken, *Idaho's Bunker Hill*, 158–59.

54. Aiken, *Idaho's Bunker Hill*, 159.

55. "Interrogatories to Bunker Hill Company," 16 September 1975, p. 8, MG 367, Box 236, Folder 4415, Bunker Hill Records.

56. "Interrogatories to Bunker Hill Company," 16 September 1975, 8.

57. Aiken, *Idaho's Bunker Hill*, 161.

58. Aiken, *Idaho's Bunker Hill*, 161.

59. "Lead Poisoning Charge Aired by Unknowns," *Kellogg Evening News*, June 29, 1961, 1.

60. "Lead Poisoning Charge Aired," 1.

61. "Lead Poisoning Charge Aired," 1.

62. "Lead Poisoning Charge Aired," 1. The issue of lead health for workers evidently still was potent in union politics a few years later. In an "Informational Bulletin" from 1964 that, among other things, blasted rival IUMMSW's record on occupational health when that union represented Bunker workers, the NWMW stated that *plumbism* was "a word that need not be looked up in the dictionary by Northwest officers or their members." The NWMW bulletin also took credit for the decrease in cases of plumbism at Bunker since that union took over, citing statistics to the effect that the number of cases had decreased from thirty-six in 1961 to sixteen in 1962, and then to four in 1963. "Informational Bulletin," Northwest Metal Workers Union, 1964, 1, in author's possession.

63. "Mine Inspector Derides Charges," *Kellogg Evening News*, June 30, 1961, 1.

64. Hays, *Beauty, Health, and Permanence*. The phrase "consumer's republic" refers to Lizabeth Cohen's highly acclaimed book *A Consumers' Republic: The Politics of Mass Consumption in Postwar America* (New York: Knopf, 2003).

65. Aiken, *Idaho's Bunker Hill*, 161.

66. "Proposed Capital Budget for 1960 Kellogg Operations of the Bunker Hill Company," MG 367, Box 3, Folder 10, Bunker Hill Records.

67. "Proposed Capital Budget for 1960 Kellogg Operations."

68. "Memorandum: Smelter Carpenter Lead Health Information," 17 June 1964, p. 1, MG 367, Box 243, Folder 4551, Bunker Hill Records.

69. "Memorandum: Smelter Carpenter Lead Health Information," 1.

70. "Memorandum: Smelter Carpenter Lead Health Information," 2.

71. "Memorandum: Smelter Carpenter Lead Health Information," 3.

72. "Memorandum: Smelter Carpenter Lead Health Information," 5.

73. "Heavy Metal Screening and Medical Program for Bunker Hill Employees," February 13, 1975, p. 5, MG 367, Box 245, Folder 4587, Bunker Hill Records.

74. "Heavy Metal Screening and Medical Program," 5.

75. Philip Landrigan to John T. Ashley, Idaho State Health Officer, 28 May 1975, p. 1, MG 413, Box 236, Folder 4424, Bunker Hill Records; Cassandra Tate, "Union Spurs Probe of Bunker Hill Smelter," *Lewiston Morning Tribune*, August 16, 1975, 1.

76. "The Carrier," March 1971, p. 1, Northwest Metal Workers Union, Piekarski papers, in author's possession.

77. "Carrier," March 1971, 1.

78. "Carrier," March 1971, 1. The author did not supplement this charge with any supporting evidence.

79. "Carrier," March 1971, 5.

80. "Heavy Metal Screening and Medical Program," 6. The highest number of screenings actually was performed in 1973, however, when 3,990 were done.

81. "Heavy Metal Screening and Medical Program," 6. The 1970 numbers of urine and blood lead readings were annualized from seven months of data.

82. "Heavy Metal Screening and Medical Program," 6.

83. "Heavy Metal Screening and Medical Program," 6.

84. James Halley, "Industrial Health Protection in Lead Exposure Areas," memo to all smelter and zinc plant employees, 24 April 1975, MG 367, Box 14, Folder 221, Bunker Hill Records.

85. Halley, "Industrial Health Protection," 1.

86. Halley, "Industrial Health Protection," 1–2. In March 1975, Bunker Hill substituted the more effective technique of blood lead testing for urine sampling. A number of smelter workers signed a petition to "strongly protest" a new rule barring the wearing of beards by smelter employees. Stating that smeltermen were not "second-class citizens," the workers went on to say that they considered the order "a brazen and unwarranted invasion of our private affairs."

87. Ken Flatt, telephone conversation with the author, March 22, 2012. The NW-MWU was replaced by the United Steelworkers of America as the bargaining agent for Bunker Hill workers in 1971. Local 7854 became the workers' new union chapter.

88. Halley, "Industrial Health Protection," 2.

89. Halley, "Industrial Health Protection," 3.

90. "Smelter Workers to Protest," *Spokesman-Review*, September 9, 1975, 9.

91. "Smelter Workers to Protest," 9.

92. Aiken, *Idaho's Bunker Hill*, 181.

93. "Smelter Workers to Protest."

94. "Smelter Workers to Protest." In some cases, though, there were positive unintended consequences of the transfers. Such was the case for Judy Etherton and three other female workers, who ended up getting to work at the Bunker Hill Mine. They were the first women ever to work in any of the mines of the Coeur d'Alenes, where the superstitions of male miners had kept them out.

95. "New Row over Workplace Sterilisation," *New Scientist*, September 25, 1980, 908.

96. "Bunker Hill Company Cited for Sterilization Policy," *Spokane Daily Chronicle*, September 13, 1980, 1; "Sterilization Rule Results in Fine," *Tri-City Herald*, September 16, 1980, 5.

97. *New Scientist*, September 25, 1980, 908.

98. "Sterilization Rule Results in Fine," 5.

99. While by the 1970s considerable scientific evidence existed of the dangers of environmental lead to fetuses, it is interesting to note that some, albeit less conclusive, evidence also existed of its dangers to the reproductive capacities of men. It does not appear, however, that Bunker Hill ever seriously considered transferring male smelter or zinc plant workers due to concerns over their abilities to father children.

100. "Sterilization Rule Results in Fine."

101. Tate, "Union Spurs Probe."

102. Halley to Woodruff, 12 December 1975, MG 367, Box 14, Folder 222, Bunker Hill Records.

103. Markowitz and Rosner, *Deceit and Denial*, 119.

104. Markowitz and Rosner, *Deceit and Denial*, 119.

105. Markowitz and Rosner, *Deceit and Denial*, 119. A watershed moment in this transformation undoubtedly was the publication of Rachel Carson's *Silent Spring* in 1962.

106. Kehoe's work at the University of Cincinnati was largely funded by the gasoline and automobile industries.

107. Markowitz and Rosner, *Deceit and Denial*, 118, 122–23. The new generation of "ecologically based" scientists tended to be funded by government.

108. The 50 µg/dl standard for workers was still somewhat higher than the 1978 standard for "undue absorption" of lead for the general population. That standard had been lowered to 30 µg/dl in 1975.

109. Markowitz and Rosner, *Deceit and Denial*, 133. By 1984, workers could return to the job only when their blood leads had fallen below 40 µg/dl.

110. "Declaration of Michael J. Wright," in *United Steelworkers of America, AFL-CIO-CLC v. Ray Marshall, Secretary of Labor, et al.*, February 12, 1980, p. 4, MG 367, Box 14, Folder 241, Bunker Hill Records.

111. "OSHA Lead Standard," 1; and "Declaration of Michael J. Wright," 10.

112. "Opposition of United Steelworkers of America to Bunker Hill Motion to Modify the Stay," in *United Steelworkers of America v. Ray Marshall, et al.*, 2.

113. "Declaration of Michael J. Wright," 4.

114. "Declaration of Michael J. Wright," 4–5.

115. "Declaration of Michael J. Wright," 5.

116. "Declaration of Michael J. Wright," 5.

117. "Declaration of Michael J. Wright," 7. In referring to "numerous control devices that have been available for some time," Wright was taking issue with the suggestion by Bunker's safety manager, Wallace Kenyon, that the company was taking all available measures to lower the blood lead levels of its smelter workers.

118. "Declaration of Michael J. Wright," 7.

119. "Declaration of Michael J. Wright," 7.

120. "Declaration of Michael J. Wright," 10.

121. "Declaration of Michael J. Wright," 12.

122. "Idaho Mine Firm Fined for Lead Dust Exposure," *Spokesman-Review*, May 4, 1979, 17.

123. "Idaho Mine Firm Fined."

124. E. Viet Howard to Secretary of Labor F. Ray Marshall and Assistant Secretary of Labor Eula Bingham, 30 January 1978, MG 367, Box 14, Folder 219, Bunker Hill Records.

125. Manning, "Bunker Hill-Government Continue Battles."

126. Robert Allen Speech in Kellogg, press release, 6 June 1978, 1, MG 367, Box 14, Folder 231, Bunker Hill Records.

127. Robert Allen Speech in Kellogg, press release, 6 June 1978, 1–2, 4.

128. E. Viet Howard to Robert Allen, "Anti-OSHA Movement by Bunker Hill Hourly Employees," 25 July 1979, MG 367, Box 14, Folder 217, Bunker Hill Records.

129. "OSHA Rapped by Employees," *Spokane Daily Chronicle*, June 5, 1979, 25.

130. "OSHA Rapped by Employees," 25.

131. "OSHA Rapped by Employees." By July, the dissident group had formed itself into an organization called Committee for Employees Against Governmental Over-regulation (CEAGO). The group appears to have been an independent, self-starting organization that was not beholden to Bunker Hill management, although management certainly applauded their efforts. CEAGO evidently had four leaders, three of whom worked at the zinc plant.

132. "OSHA Not Opposed," *Lewiston Tribune*, June 8, 1979.

133. "Bunker Hill Company Cited," 1.

134. "Negotiating with Bunker," *Spokesman-Review*, September 17, 1981, 8.

135. Joann S. Lublin, "Union Wins OSHA Concessions to Save Idaho Smelter Despite Health Risks," *Wall Street Journal*, September 25, 1981, 14.

136. Lublin, "Union Wins OSHA Concessions," 14.

137. Lublin, "Union Wins OSHA Concessions," 14.

138. Lublin, "Union Wins OSHA Concessions."

139. "Summary" (of NIOSH study), p. 1, MG 413, Box 247, Folder 4633, Bunker Hill Records, Special Collections.

140. "Summary" (of NIOSH study).

141. "Summary" (of NIOSH study), 1.

142. "Summary" (of NIOSH study), 1. Mean air levels were 1,850 for zinc and 18 $\mu g/m^3$ for arsenic.

143. "Summary" (of NIOSH study), 1. Their mean cadmium exposure factored in at 521 $\mu g/m^3$.

144. "Summary" (of NIOSH study), 1.

145. "Summary" (of NIOSH study), 1.

146. "Summary" (of NIOSH study), 1. This of course was reminiscent of the Shoshone Project's findings about children with high blood leads in the wake of the baghouse fire.

147. "Summary" (of NIOSH study), 2.

148. "Summary" (of NIOSH study), 2. Medical evidence shows that those who have suffered lead intoxication sufficient to require chelation therapy are thereafter more susceptible than others to lead intoxication.

149. S. G. Selevan, P. J. Landrigan, F. B. Stern, and J. H. Jones, "Mortality of Lead Smelter Workers," *American Journal of Epidemiology* 122, no. 4 (1985): 673–83.

150. Selevan et al., "Mortality of Lead Smelter Workers," 677.

151. Selevan et al., "Mortality of Lead Smelter Workers," 677. The Bunker Hill smelter workers' population was considered "white" for the purposes of this study. Although racial demographic data could not be obtained for all the workers, based upon the fact that in 1976, 96 percent of the smelter's workers were white, the investigators felt safe in making this assumption. *Standard mortality ratio* refers to the ratio between the observed number of deaths in a study population and the number of deaths that would be expected in a standard population.

152. Selevan et al., "Mortality of Lead Smelter Workers," 682.

153. Selevan et al., "Mortality of Lead Smelter Workers," 680.

154. Selevan et al., "Mortality of Lead Smelter Workers," 677.

155. Selevan et al., "Mortality of Lead Smelter Workers," 681.

156. Selevan et al., "Mortality of Lead Smelter Workers," 680. Six of these deaths occurred in the Sunshine Mining disaster.

157. Cynthia Taggart, "Bunker Hill Men Still at Risk from Lead," *Spokesman-Review*, May 21, 1994.

158. McEvoy, "Working Environments," S146–47.

159. Hays, *Beauty, Health, and Permanence*, 250–53, 374–75. Such a facility likely also would have operated far more efficiently as well and produced greater profits for the company. While natural resource companies such as Bunker Hill like to frame debates over environmental regulation in terms of "imposed costs" to them that presumably make them less healthy and wealthy businesses, companies that get ahead of the curve and invest in new, cleaner technology can reap both environmental and business rewards.

160. Beck, *Risk Society*, 23.

161. This is certainly not to minimize the health effects of working for Bunker Hill on miners who suffered high rates of accident and also were beset with serious health problems including silicosis, nor that on community members, particularly those who lived nearby the smelter, and thus resided in an environment that was heavily impregnated with lead and other heavy metals.

162. Taggart, "Bunker Hill Men Still at Risk."

Chapter 7: A Brave New World

Epigraph: Beck, *Risk Society*, 39.

1. "Bunker Hill: Bush Aid Enlisted," *Spokesman-Review*, September 6, 1981, A13. Such environmental regulations were viewed by many as a significant roadblock to a sale of the company.

2. "Bunker Hill: Congressmen Brief President on Problems," *Lewiston Tribune*, September 22, 1981. According to Craig, Reagan "showed real alarm" over Bunker's impending closure.

3. "Negotiating with Bunker," *Spokesman-Review*, September 17, 1981, 8.

4. "Negotiating with Bunker," 8.

5. Frank Bartel, "Bunker Hill Closure to Cost Spokane 800 Jobs," *Spokane Chronicle*, November 19, 1981, 44.

6. Bartel, "Bunker Hill Closure," 44.

7. "Bunker Union Recommends Rejection of Wage Cuts," *Kellogg Evening News*, January 14, 1982, 1. Duane Hagadone, a leader of the investor consortium that eventually materialized as a possible Bunker buyer, estimated that a more accurate forecast of losses in 1982 was $54 million.

8. Jim Fisher, "Bunker Hill Purchase Option Raises Hopes," *Kellogg Evening News*, December 11, 1981, 1.

9. Sue Lancaster, "Cautious Elation Greets Announcement," *Kellogg Evening News*, December 11, 1981, 1.

10. Doug Barker, "Union Members to Review New Labor Agreement," *Kellogg Evening News*, January 15, 1981, 1. The group, which offered Gulf $50 million in cash and $15 million in payments from future profits, also planned to trim the company's workforce by five hundred in the coming year.

11. "Bunker Hill: Bush Aid Enlisted," A13.

12. Jim Fisher, "Simplot Says 'Yes'; Unions to Get Proposal," *Kellogg Evening News*, January 12, 1982.

13. "Bunker Union Recommends Rejection of Wage Cuts," 1.

14. Doug Barker, "Bunker Pact in Limbo after Vote, Resignations," *Kellogg Evening News*, 1. Despite the International's ruling, Ken Flatt signed the deal, believing the January 17 vote represented the will of his Local's membership. Flatt's signature, however, of course did not override the International's decision and the January 17 vote remained unofficial. According to Flatt, he felt strongly "ambivalent" about the International's stance on the vote, believing that while it acted against the expressed will of a majority of his membership, it also was a necessary action to prevent an unending round of lawsuits against the International by Local 7854 members aggrieved by the loss of retirement benefits that the proposed contract entailed. Ken Flatt, telephone conversation with the author, March 22, 2012.

15. Flatt, telephone conversation with the author, March 22, 2012.

16. Jim Fisher, "Steelworkers Remove Officers, Elect New Leaders," *Kellogg Evening News*, 1.

17. Flatt, telephone conversation with the author, March 22, 2012. Shortly after the vote, the Ron Byrd faction occupied the Local 7854 union hall in Kellogg. Following the "occupation," Flatt was stopped on the highway outside of town on his way to work and told by a Kellogg police officer to return home, apparently because the officer feared a violent confrontation at the hall. Flatt would not return to work at the Steelworkers Building for approximately a year.

18. Jim Fisher, "Petitions Protesting Union Pact Refusal Sent to Reagan by Salaried Employees," *Kellogg Evening News*, January 28, 1982, 1.

19. "Rescue Effort Continues," *Kellogg Evening News*, January 22, 1982, 1.

20. "Symms and Craig Meet with Union Chief," *Kellogg Evening News*, February 2, 1982, 1; "McBride Rejects Plea to Save Pact," *Kellogg Evening News*, February 3, 1982, 1.

21. Fisher, "Petitions Protesting Union Pact Refusal," 1; Jim Fisher, "Steelworkers Remove Officers, Elect New Leaders," *Kellogg Evening News*, 1.

22. Doug Barker, "Bunker Hill Sells Sinter Machine, No Rescue Seen," *Kellogg Evening News*, February 15, 1982, 1. Although many in the Silver Valley blamed the International Steelworkers' stance for the failure to keep Bunker Hill running, it must be noted that the bid to keep the company, and particularly its aged smelter, going always faced steep odds. Evidence of this came in the fact that a smaller version of the Simplot group purchased Bunker Hill in November 1981, but chose not to reopen the mine for several years due to low silver prices, and never reopened the lead or zinc plants.

23. "Devil's bargain" is a reference to Hal K. Rothman's history of tourism in the twentieth-century American West. Rothman finds that the choice of tourism by communities as an economic strategy "typically fails to meet . . . expectations. . . . Tourism transforms culture into something new and foreign; it may or may not rescue economies." Hal K. Rothman, *Devil's Bargains: Tourism in the Twentieth-Century American West* (Lawrence: University Press of Kansas, 1998), 10.

24. Particularly affected are those involved in such "dirty industries" as mining, smelting, steel, oil, chemicals, and nuclear or chemical weapons.

25. Berman, *All That Is Solid*, 347.

26. Jim Fisher, "Unskilled Workers Bunker Hill's Main Concern, Kendrick Says," *Kellogg Evening News*, December 2, 1981, 1. Decent economic conditions, however, hardly existed at the time, as Kendrick admitted.

27. "Bunker Workers Face Tough Job Hunt," *Spokesman-Review*, December 9, 1981, A7.

28. "Bunker Workers Face Tough Job Hunt."

29. Sue Lancaster, "Evans: Gulf Won't 'Walk Away' without Leaving Help," *Kellogg Evening News*, December 1, 1981, 1.

30. "Aid Aimed at Kellogg Jobless," *Spokane Chronicle*, March 31, 1982, 10. The benefits issued from a federal program designed to assist US workers forced out of jobs by the pressure of foreign competition. Only Bunker's former zinc workers were deemed eligible.

31. "Silver Valley Responds," editorial, *Spokesman-Review*, January 19, 1983, A4.

32. Steven Pike, "Silver Valley Schools Counting on Emergency State Funding," *Spokesman-Review*, March 4, 1983. The projected decrease in revenues, in turn, was caused by a 60 percent loss in property tax assessments over the previous two years. This was due in large measure to the loss of much of Bunker Hill's plant and equipment from the school district's tax base.

33. "Wallace Schools Seek Economic Lifeline," *Lewiston Morning Tribune*, August 5, 1983, 3D.

34. "Schools Face Cuts if Budget Not Hiked," *Spokesman-Review*, May 9, 1983.

35. "Wallace Schools Seek Economic Lifeline."

36. "Schools Face Cuts."

37. "Table 2: Counties—Employees, Payroll, and Establishments, by Industry: 1981," in US Bureau of the Census, *County Business Patterns, 1980, Idaho: Employment and Payrolls, Number and Employment, Size of Establishment, by Detailed Industry* (Washington, DC: US Department of Commerce, Bureau of the Census, 1982), 44. In tabulating employment data, the census bureau does not include government or railroad employees or the self-employed.

38. "Table 2: Counties—Employees, Payroll, and Establishments, by Industry: 1981."

39. I am using the term "service industry" loosely here, not just to refer to the hostelry and recreation industries, but to all those industries that essentially exist to provide services to basic industry (e.g., law, banking, real estate, transportation, utilities).

40. Bill Morlin, "Railroads Hurry to Catch Up," *Spokesman-Review*, July 31, 1983, C1.

41. Morlin, "Railroads Hurry to Catch Up," C1.

42. "Table 2: Counties—Employees, Payroll, and Establishments, by Industry: 1987," in US Bureau of the Census, *County Business Patterns, 1980, Idaho: Employment and Payrolls, Number and Employment, Size of Establishment, by Detailed Industry* (Washington, DC), 54. In many ways the picture actually was worse by 1987, the economy more hollowed out. Annualized March 1987 US Census figures showed only 2,530 people employed in Shoshone County, down from 3,532 in 1983, and payrolls of $9,581,000, slightly over half of 1983's $17,622,000. By this time, far more so than in 1983, the area's old industrial economy had really faded and had yet to be meaningfully replaced by tourism or revenues from the EPA Superfund cleanup.

43. David Bond, "Hope and Fear in Silver Valley," *Spokesman-Review*, June 21, 1987, 1.

44. David Bond, "Tramp Miners Try, Try Again," *Spokesman-Review*, June 22, 1987, A5. At the time, both El Paso and Trail were the sites of lead smelters. At this writing the Trail smelter is still operating while the El Paso smelter has long since been shuttered.

45. Bond, "Tramp Miners Try, Try Again," A5. The Bunker Hill shutdown was a big part of the employment problem in the valley during the 1980s but certainly not the entire story. Big mines in the district such as the Sunshine, the Lucky Friday, and the Galena either shut down completely or were open only intermittently during the decade, largely in response to highly fluctuating metal prices. Regarding the lower pay outside the Coeur d'Alenes, whereas miners as the Sunshine Mine made about twenty-five dollars an hour in 1987, those at the Stillwater Mine in Nye, Montana, earned about thirteen dollars an hour.

46. Bond, "Tramp Miners Try, Try Again," A5.

47. Doug Barker, "Silver Valley Residents 'Brainstorm' to Rebuild Area's Economy," *Kellogg Evening News*, February 9, 1982, 3. The idea of playing on Kellogg's western and mining heritage in the creation of a theme town was given serious consideration, but city leaders eventually decided to cede that field to Wallace, its neighbor eleven miles to the east. While Kellogg's architecture lacked a common theme, Wallace's revolved around its many interesting, turn-of-the-century redbrick buildings located in the town's compact business district. Its architecture made Wallace something of a natural to pursue the "historic mining town" theme. In a unique example, the entire town eventually was placed on the National Historic Register. Among other historic attractions, Wallace boasts the Northern Pacific Depot Railroad Museum, the Wallace District Mining Museum, and the Oasis Bordello Museum.

48. "Gondola Scheme Gaining Support," *Spokane Chronicle*, June 22, 1985, 10.

49. L. M. Griffin to R. L. Haffner, 4 November 1969, MG 367, Box 98, Folder 2231, Bunker Hill Records. In addition to leasing the land at no charge, Bunker also donated at least $70,000 in money and services to Fabulous Valley, and owned a little less than one-sixth of the company's shares.

50. In the fall of 1982, the Bunker Limited Partnership, Inc., owned by Jack Simplot, Duane Hagadone, Harry F. Magnuson, and Jack Kendrick, purchased ownership of the remaining former Bunker Hill properties. These included the Bunker Hill Mine and Silverhorn. The consortium ran the ski resort for one season before deciding to sell the business to the City of Kellogg. In 1987, they reopened the Bunker Hill Mine and ran it with a reduced, nonunion crew into the early 1990s, when they went out of business.

51. "Kellogg Facing Need to Diversify Economy," *Spokesman-Review*, June 24, 1986, 4.

52. "City to Get Bavarian Look," *Bulletin* (Bend, OR), December 27, 1985, A3; Timothy Egan, "Resort or Mining Town? Kellogg Makes a Change," *Harlan Daily Enterprise* (Harlan, KY), July 14, 1989, 8. Under the ordinance, existing signage was "grandfathered in" and allowed to remain as it was until 1993.

53. "Big Gondola Funding Goal for Kellogg," *Spokane Chronicle*, September 10, 1986, 3.

54. David Bond, "Federal Aid for Gondola Is Uncertain," *Spokane Chronicle*, June 20, 1986, 1.

55. "Idaho Town Wants Lift from Taxpayers," *Vindicator* (Youngstown, OH), February 1, 1988, 3. Though a staunch Republican with strong conservative credentials, McClure was skewered in the press for securing this type of funding for the Kellogg gondola. *Newsweek* labeled the project "ridiculous"; national radio personality Paul

Harvey blasted it on his show; conservative columnist James J. Kilpatrick hammered it; the national group Citizens Against Public Waste was critical; and President Ronald Reagan singled the gondola appropriation out, calling for Congress to rescind it and labeling the grant "a prime example of the use of federal tax dollars to finance a local project which benefits a very limited area." For the Reagan and Kilpatrick critiques, see James J. Kilpatrick, "Of Cops, Games and Gondolas," *Lewiston (ME) Daily Sun*, March 29, 1988, 4.

56. "Gondola Firm Offers Hope for Kellogg," *Spokesman-Review*, September 2, 1988, 1.

57. "Gondola Firm Offers Hope," 1.

58. "Destination North Idaho," *Spokane Chronicle*, June 29, 1990. Silverhorn was closed for the 1989–1990 season, while the gondola was being built. About $200,000 was put into improving the resort and it was reopened with its new name, under Von-Roll's management, on June 30, 1990.

59. "Destination North Idaho."

60. Cary Ordway, "New Silver Mountain Ski Area Will Be a First Class Resort," *Spokane Chronicle*, January 5, 1990, 10. Perhaps fittingly, the primary means of production for Kellogg's new, leisure economy, the gondola's wheelhouse, was close to the cornerstone means of production for Kellogg's former mining and smelting economy.

61. "Ski Resorts by Vertical Feet," accessed October 14, 2011, http://verticalfeet.com/; Slackpacker, "Colorado Ski Area Guide—Very Complete," accessed October 14, 2011; http://www.skiernet.com/ski_co.html. In 2011, Silver Mountain has expanded to 2,202 vertical feet and 1,500 acres of skiable terrain, served by six lifts and the gondola. Despite its expansion, the resort remains a mid- to upper-mid-sized ski area, trailing far behind such destination resorts as Telluride (4,425 feet, 1,700 acres), Vail (3,330 feet, 5,289 acres), and Park City (3,100 feet, 2,200 acres) and even lagging behind nearby Schweitzer Mountain (2,400 feet, 2,900 acres) and northwest Montana's Whitefish Mountain Resort (2,300 feet, 3,000 acres).

62. Ordway, "New Silver Mountain Ski Area." Hagadone Hospitality was owned by Duane Hagadone. In 1987, the company opened the large, luxury Coeur d'Alene Resort on the shores of Lake Coeur d'Alene. Since his involvement in the failed effort to purchase the Bunker Hill Company and keep it running in 1981–1982, Hagadone had become heavily involved in Silver Valley affairs. In addition to his interest in Silver Mountain, as a member of the ownership group Bunker Limited Partnership, Hagadone controlled the defunct Bunker Hill Company's remaining holdings in the Kellogg area. In 1985 he purchased the *Kellogg Evening News*, renaming it the *Shoshone News-Press*. As of 2015, the newspaper is the only daily operating in the Silver Valley.

63. Ordway, "New Silver Mountain Ski Area."

64. Steve Massey, "Skier Turnout Forces Resort to Make Room," *Spokesman-Review*, December 12, 1990, A12. Forsch's prediction proved to be prescient, although the boom he anticipated would not arrive for a good many years and then proved to be relatively short-lived.

65. Steve Massey, "$5.3 Million Will Improve Ski Resort," *Spokesman-Review*, September 13, 1990, A3.

66. Egan, "Resort or Mining Town?," 14.

67. "Kellogg's Face Still in Need of a Lift," *Spokesman-Review*, May 22, 1989, A6.

68. "Kellogg's Face Still in Need," A6.

69. Egan, "Resort or Mining Town?"

70. "Kellogg Torn by Campaign," *Lewiston Morning Tribune*, January 3, 1993, 6C.

71. J. Todd Foster, "Bavarian Motif for Store Signs Bugs Merchants," *Spokesman-Review*, December 16, 1992, A1.

72. "Kellogg Torn by Campaign," 6C.

73. Massey, "Skier Turnout," A12; Steve Massey, "Silver Valley Launches $1 Million Expansion," *Spokesman-Review*, August 30, 1991, A12. "Skier visits" is the terminology employed by the snow sports industry to catalogue the number of lift tickets sold at a given ski resort.

74. PRN Newswire, "Winter Sports, Inc. Seeks to Acquire Silver Mountain Ski Resort," April 8, 1996, accessed October 14, 2011, http://www.thefreelibrary.com/WINTER+SPORTS,+INC.+SEEKS+TO+ACQUIRE+SILVER+MOUNTAIN+SKI+RESORT-a018171033. Big Mountain has since been renamed. The Whitefish, Montana, resort is now called the Whitefish Mountain Resort.

75. Becky Kramer, "Ski Season Verges on Legendary," *Spokesman-Review*, April 13, 1999, A8; "Telluride's Formula for Success," *New York Times*, November 17, 1991.

76. Eric Torbenson, "Details of Ski Hill Plan Still Lacking," *Spokesman-Review*, May 31, 1996, A10.

77. Silver Mountain Resort, "History of Silver Mountain Ski Resort," accessed October 14, 2011, http://www.silvermt.com/site_Our-Resort/OR-Default.aspx?page=OR-History.

78. Silver Mountain Resort, "History of Silver Mountain Ski Resort."

79. Becky Kramer, "Wal-Mart Considers Store in Silver Valley," *Spokesman-Review*, April 7, 2005, A8.

80. Becky Kramer, "Silver Turns to Gold: Condominium Surge Gives Silver Valley an Economic Lift," *Spokesman-Review*, February 4, 2007.

81. Kramer, "Silver Turns to Gold."

82. William Yardley, "In High Prices, Moribund Mines Find Silver Bullet," *New York Times*, April 3, 2008.

83. Silver Mountain Resort, "History of Silver Mountain Ski Resort."

84. I use the term "old West" here to contrast the Rocky Mountain West's traditional natural resource extraction economy, with all its benefits and harms, with an idea popularized over the past twenty-plus years of a "new West," in which a tourist and second-home economy replete with cappuccino bars, ski slopes, "blue ribbon" trout streams, and mountain biking trails is often hailed as a preferable, more ecological way of life for westerners. While it may be somewhat "cleaner," ecologically speaking, than were its "old West" alternatives, it is also true that multinational corporations often have turned to former Third World countries to harvest raw materials in ecologically unfriendly ways, while many "new West" communities have been hollowed out economically and demographically, as wages in the tourist economy drop precipitously and longtime community residents can no longer afford to live in homes whose property taxes have risen to reflect new market prices.

85. Yardley, "In High Prices." According to 2006 economic data cited in the article, the miners' average wage of $57,000 was more than double the average for the county's other jobs.

86. Idaho Department of Environmental Quality, "Records of Decision," accessed October 17, 2011, http://www.deq.idaho.gov/regional-offices-issues/coeur-d'alene/bunker-hill-superfund-site/records-of-decision.aspx. At the time the Bunker Hill site was the largest in the Superfund system. It has since fallen into second place in this dubious category. In the mid-1980s, the EPA gave serious consideration to the idea of physically moving the entire town of Kellogg due to the fact that the community had been built atop mine tailings and stood to be so difficult to clean up. The idea eventually was dropped because of negative reaction from the community.

87. "Bulldozers Signal the Cleanup of Contaminated Community," *Eugene (OR) Register-Guard*, January 13, 1989, 6C; Timothy Egan, "Some Say Mining Company's Move Could Thwart U.S. Plan for Cleanup," *New York Times*, October 2, 1990. The EPA's inspector general wrote in 1990 that Russell "took extraordinary steps to prevent formal enforcement actions from being initiated against the owners" of the Bunker Hill mining site. It remains unclear why this occurred. Russell's critics, however, alleged that the Idaho native was trying to protect local businesses.

88. Idaho Department of Environmental Quality, "Records of Decision," accessed October 17, 2011, http://www.deq.idaho.gov/regional-offices-issues/coeur-d'alene/bunker-hill-superfund-site/records-of-decision.aspx.

89. Idaho Department of Environmental Quality, "Records of Decision."

90. Rich Roesler, "Move Afoot to Save Valley Smokestack," *Spokesman-Review*, June 29, 1995, A1.

91. Karen Dorn Steele, "EPA Strikes Vein of Anger," *Spokesman-Review*, July 21, 2002.

92. Steele, "EPA Strikes Vein of Anger"; "EPA Plan for Basin Unveiled," *Spokesman-Review*, September 13, 2002. This proposed expenditure came atop the $253 million that already had been spent on the Bunker Hill Superfund site. Because Gulf Resources went bankrupt in 1993, the bulk of these funds had to be drawn from the Superfund trust fund.

93. Karen Dorn Steele, "EPA Cleanup Plan Spreads into Basin," *Spokesman-Review*, October 21, 2001.

94. Steele, "EPA Strikes Vein of Anger."

95. Steele, "EPA Strikes Vein of Anger."

96. Steele, "EPA Cleanup Plan Spreads into Basin." The columnist, David Bond, a reporter who, prior to his service with the *Shoshone News-Press* and his adoption of an anti-EPA advocacy position, already possessed years of experience covering the Silver Valley for the *Spokesman-Review*, claimed he only made the comment in jest.

97. Nicholas K. Geranios, "EPA Confronts Silver Valley Cleanup Plan," *Beaver County (PA) Times*, October 23, 2001, B10.

98. Cynthia Taggart, "Kellogg Activist Honored Nationally," *Spokesman-Review*, September 7, 2001, 1.

99. Taggart, "Kellogg Activist Honored Nationally."

100. Cynthia Taggart, "Activist Wins Gold for Getting Lead Out," *Spokesman-Review*, September 20, 2001, B1.

101. Taggart, "Activist Wins Gold," B1.

102. Taggart, "Activist Wins Gold," B1.

103. Nick Rotuno, "Cleanup Moving Forward," *Shoshone News-Press*, February 22, 2010.

104. Nicholas K. Geranios, "Mining Co. to Pay $263M to Settle Superfund Suit," *Bloomberg Businessweek*, June 13, 2011, accessed October 18, 2011, http://www.businessweek.com/ap/financialnews/D9NR5ODO0.htm. The remaining $66 million was slated to go to the Coeur d'Alene Indian Tribe, the State of Idaho, and other federal agencies, for use in their own cleanup efforts in the basin.

105. Dick Harwood, "EPA Editorial," *Shoshone News-Press*, October 7, 2010; Nicole Nolan, "Residents Blast EPA Proposal," *Shoshone News-Press*, October 21, 2010.

106. US Environmental Protection Agency, Region 10: The Pacific Northwest, "ROD Amendment for Upper Basin and Box," *Bunker Hill Mining and Metallurgical Superfund Site (Coeur d'Alene Basin)*, accessed October 18, 2011, http://yosemite.epa.gov/R10/CLEANUP.NSF/sites/bh+rod+amendment.

107. Dan Drewry, "Stop the EPA Steamroller," editorial, *Shoshone News-Press*, November 17, 2010.

108. Nicole Nolan, "Residents Blast EPA Proposal," *Shoshone News-Press*, October 21, 2010.

109. Nolan, "Residents Blast EPA Proposal." John Magnuson, son of the late Wallace mining magnate Harry F. Magnuson, opined that the sessions should be renamed "talking sessions" for this reason.

110. Harwood, "EPA Editorial."

111. Drewry, "Stop the EPA Steamroller."

112. Alecia Warren, "EPA Funding Will Help Assess Cleanup Sites," *Shoshone News-Press*, May 19, 2011. The proposed savings would come from dropping plans to line the bottom of a long stretch of the South Fork of the Coeur d'Alene River with a plastic liner. Instead, the agency would install much less expensive groundwater treatment.

113. Karen Dorn Steele, "Superfund's Silver Lining," *Spokesman-Review*, July 28, 2002.

114. Steele, "Superfund's Silver Lining."

115. Steele, "Superfund's Silver Lining."

116. Steele, "Superfund's Silver Lining."

117. Nolan, "Residents Blast EPA Proposal." At the public hearing, State Representative Mary Lou Shepherd (D-Wallace) took issue with the EPA's contention that it was helping to sustain the local economy by providing jobs. Representative Shepherd countered that the agency actually was hampering economic development because its jobs were only seasonal (summer). Interestingly, I have not run across any complaints about the seasonality (winter) of the bulk of Silver Mountain's employment.

118. Greg Marsh, "Route of the Hiawatha: Mountain Bike Trail," accessed October 19, 2011, http://wallace-id.com/skilookout/taft.html.

119. Nicole Nolan, "A Record for Route of the Hiawatha," *Shoshone News-Press*, September 7, 2010.

120. Becky Kramer, "Portions of CdA Trail Closed by Flood Damage," *Spokesman-Review*, June 13, 2008.

121. Karen Dorn Steele, "Bike Path Cleanup Assailed," *Spokesman-Review*, October 13, 2002; and personal experience of riding the trail, September 2008.

122. Steele, "Bike Path Cleanup Assailed." Under its agreement, the Union Pacific maintains perpetual responsibility for maintenance of the trail. The State of Idaho and the Coeur d'Alene Indian Tribe perform this maintenance.

123. Becky Kramer, "Silver Valley Trail Bringing in Gold," *Spokesman-Review*, July 21, 2005.

124. Kramer, "Silver Valley Trail Bringing in Gold."

125. Idaho Department of Labor, "Shoshone County Work Force Trends, September, 2011," accessed October 19, 2011, http://labor.idaho.gov/publications/lmi/pubs/ShoshoneProfile.pdf.

126. Idaho Department of Labor, "Shoshone County Work Force Trends, September, 2011."

127. Idaho Department of Labor, "Shoshone County Work Force Trends, September, 2011." The unemployment rate was 15.9 percent.

128. Idaho Department of Labor, "Shoshone County Work Force Trends, September, 2011." The Lucky Friday is near Mullan and the Galena is near Wallace. Both are owned by Coeur d'Alene mining companies, the Lucky Friday by the Hecla Mining Company and the Galena by Coeur Mines.

129. Becky Kramer, "Mine Death Put on Bosses," *Spokesman-Review*, September 17, 2011.

130. Kramer, "Mine Death Put on Bosses," and Sarah Rose Fredlund, "Dealership Honored Again," *Shoshone News-Press*, January 7, 2010.

131. Idaho Department of Labor, "Shoshone County Work Force Trends, September, 2011," and CoverageforAll.org: Foundation for Health Coverage Education, "2011 Federal Poverty Level," accessed October 19, 2011, http://www.coverageforall.org/pdf/FHCE_FedPovertyLevel.pdf.

132. Nicole Nolan, "Uptown Kellogg Ideas Sought," *Shoshone News-Press*, September 10, 2010. Among the long-boarded-up Uptown structures is the big, redbrick YMCA building, Uncle Bunker's old pride and joy. In addition to the heightened signs of urban blight in Uptown, the City of Kellogg experienced a dramatic uptick in crime rates during 2009.

133. Nolan, "Uptown Kellogg Ideas Sought."

134. Nolan, "Worried about Uptown," *Shoshone News-Press*, September 15, 2010.

135. Nolan, "Worried about Uptown." In making her case, Stinson cited the recent return to I-90 of a billboard sign that for years had welcomed highway visitors to Kellogg. The sign, bearing a large depiction of a jackass, read "You are now nearing KELLOGG, the town which was Discovered by a JACKASS—and which is inhabited by its Descendants."

136. Brenda Stinson, "Improved Signs for Visitors Is a Must for Kellogg," editorial, *Shoshone News-Press*, November 9, 2010. Numerous process improvements were suggested, however, such as improving street and highway signage to better direct visitors to local businesses, doing more grant writing, and establishing a Kellogg Facebook page.

Conclusion

1. Long, *Production and Disposal of Mill Tailings*, 2. This figure is measured in 1997 dollars.

2. Since the early 2000s, with the enlargement of the Superfund site to include a significant area downstream from the original twenty square-mile "box" centered on the old Bunker Hill smelter, the site has grown well beyond the geographic boundaries of the Coeur d'Alenes.

3. Long, *Production and Disposal of Mill Tailings*, 2; Coeur d'Alene River and Lake Commission, *Report and Recommendations*, 50.

4. US Environmental Protection Agency, Region 10: The Pacific Northwest, "Cleanup Work: Site Background," *Bunker Hill Mining and Metallurgical Superfund Site (Coeur d'Alene Basin)*, accessed April 9, 2012, http://yosemite.epa.gov/R10/CLEANUP.NSF/bh/Cleanup+Work.

5. Drumheller and Welsh, "Children Left in the Dust," A1. The study interviewed 917 former Silver Valley children and tested the nervous systems and bone lead of 281 of them. For a control group, the study used comparable numbers of Spokane residents of roughly the same age.

6. "Summary" (of NIOSH study), p. 1, MG 413, Box 247, Folder 4633, Bunker Hill Records.

7. My use of the "biological core" concept owes to the work of environmental and legal historian Arthur McEvoy, and the term *workscape* is drawn from the work of environmental historian Thomas Andrews.

8. Although Hecla and ASARCO, two companies long involved in mining and milling in the district, recently agreed to large payouts for remediation of the Coeur d'Alene basin, the bankruptcy of Gulf Resources in the early 1990s left the US taxpayer to foot the bill for Gulf-Bunker's significant share of the cleanup costs for the northern Idaho and eastern Washington environment. Although former Gulf-Bunker employees receive contractually obligated health insurance benefits from their former employer, it appears no direct payout has been made to former smelter employees who have suffered renal disease or other health problems contracted while working for the company. In any event, it is doubtful that any monetary recompense can ever fully compensate an individual for lost health or completely restore a badly fouled environment.

9. Arthur F. McEvoy, *The Fisherman's Problem: Ecology and Law in the California Fisheries, 1850–1980* (Cambridge, UK: Cambridge University Press, 1986), 9–13. The fisherman's problem, as employed by McEvoy, refers to the tendency of individuals to overharvest natural resources (e.g., fish) that are not privately owned because the

social costs of the unsustainable gathering of such resources are borne primarily by society in general while the gains to be made are realized individually.

10. "Declaration of Michael J. Wright," and Hooper, testimony, 18 August 1975, 9.

11. US Bureau of Fisheries, *Pollution of the Coeur d'Alene*, 55.

12. It is unclear that the "Kimberly system" would have cost more. If, however, the MOA already had gone ahead and begun to install the dredge, the mine owners may well have deemed the additional cost of the flume system too great an expense.

13. Wirth, *Smelter Smoke in North America*, 82–83.

14. The company did, however, bow to these pressures to the extent of constructing a sulfuric acid plant in the early 1950s. This constituted a lower cost, less onerous, and significantly less effective means of addressing its "sulfur problem."

15. Aiken, *Idaho's Bunker Hill*, 37.

16. Among the capitalists who were "present at the creation" of the mining district were Simeon Reed, Cyrus McCormick, D. O. Mills, and William H. Crocker. Among the largest corporations were the Federal Mining and Smelting Company, which eventually was controlled by the Guggenheims, and Bunker Hill. John Hays Hammond was among the famous mining engineers who plied their trade in the Coeur d'Alenes during this period.

17. This strife culminated in the hijacking of a train in Burke in 1899 by unionists and the blowing up of the Bunker Hill mill in Kellogg. Following that, union (WFM) leaders and their suspected supporters were rounded up and put in stockades ("bull-pens") in Kellogg, in some cases for months.

18. To paraphrase Marx and Engels.

Epilogue

1. Michael Wines, "Lead Sickens 1,300 Children in China," *New York Times*, August 21, 2009.

2. Sky Canaves, "Beijing Says New Measures Needed to Fight Lead Poisoning," *Wall Street Journal*, September 4, 2009.

3. Canaves, "Beijing Says New Measures Needed."

4. Canaves, "Beijing Says New Measures Needed."

5. Wines, "Lead Sickens 1,300 Children in China."

6. All of these facilities smelted ore concentrates.

7. "Lead Poisoning Haunts Chinese Smelter Communities," *New York Times*, August 26, 2009.

GLOSSARY

A VOCABULARY OF COEUR D'ALENES' MINE, MILL AND SMELTER WASTES

Location, Phase of Production	Intended Product	By-product, Waste
Mine	Ore	Waste Rock
Mill, Concentrator (flotation)	Concentrates	Tailings: Sands or Slimes
Smelter	Lead Bullion	SO_2, Fumes, Smoke, Slag
Electrolytic Refinery	99.99% Refined Zinc	SO_2

Terminology

Concentrates: The desired mineral that is left after impurities have been removed from mined ore as part of the milling process.

Fume: An often irritating or offensive smoke, vapor, or gas produced through the smelting process.

Mill: A unit operation designed to grind or break ore into smaller pieces.

Ore: A naturally occurring mineral that contains a valuable constituent, such as metal, for which it is mined and worked.

Sands: The larger waste materials (i.e., tailings) remaining after the milling of ores has been completed.

Slag: The stony waste matter separated from metals during the smelting or refining of ore.

Slimes: A product of wet crushing consisting of ore ground so fine as to pass a 200-mesh screen, waste from the milling process.

Smelter: An industrial facility whose purpose is to produce a metal from an ore.

Tailings: The waste material (i.e., sands or slimes) left over after milling or concentrating ore.

SELECTED BIBLIOGRAPHY

Manuscript Collections

Bunker Hill Company Photographic Collection. In possession of Kellogg Public Library.

Local No. 18, International Union of Mine, Mill and Smelter Workers, Northwest Metal Worker Union, Local 7854 of the United Steelworkers of America. In possession of Katherine Aiken.

National Archives and Records Administration, Pacific Alaska Region (Seattle)

Records Group 95, US Forest Service, Coeur d'Alene National, Historical Collection Circa 1905–1990

University of Idaho Special Collections

Manuscript Group 130. Bunker Hill correspondence regarding some of its smaller mines, 1894–1937.

Manuscript Group 187. Records of the Bunker Hill Research and Development Division.

Manuscript Group 188. Bunker Hill legal cases, 1899–1902.

Manuscript Group 367. Bunker Hill Mining Company records. 1887–1984.

Manuscript Group 413. Bunker Hill Company-Pintlar Corporation Records, 1968–1993. This collection principally covers the period from Gulf Resources and Chemical Corporation's takeover of Bunker Hill in 1968 through the bankruptcy of Pintlar Corporation, Bunker Hill's final owner, in 1993.

Sources

Aiken, Katherine G. *Idaho's Bunker Hill: The Rise and Fall of a Great Mining Company, 1885–1981.* Norman: University of Oklahoma Press, 2005.

Aiken, Katherine G. "'Not Long Ago a Smoking Chimney Was a Sign of Prosperity': Corporate and Community Response to Pollution at the Bunker Hill Smelter in Kellogg, Idaho." *Environmental History Review* 18, no. 2 (1994): 67–86.

Amundson, Michael A. "Yellowcake to Singletrack: Culture, Community and Identity in Moab, Utah." In Nicholas, Bapis, and Harvey, *Imagining the Big Open,* 151–62.

Amundson, Michael A. *Yellowcake Towns: Uranium Mining Communities in the American West.* Boulder: University Press of Colorado, 2002.

Andrews, Thomas G. *Killing for Coal: America's Deadliest Labor War.* Cambridge, MA: Harvard University Press, 2008.

Arrington, Leonard J. *History of Idaho.* Vol. 2. Moscow, ID: University of Idaho Press, 1994.

Beck, Ulrich. *Risk Society: Towards a New Modernity.* London: Sage, 1992.

Beck, Ulrich. *World at Risk.* Cambridge, UK: Polity, 2009.

Bell, Robert N. "Coeur d'Alene Items, June, 1917." *Engineering and Mining Journal*, July 7, 1917, 6.

Berman, Marshall. *All That Is Solid Melts into Air: The Experience of Modernity*. New York: Simon and Schuster, 1982.

Blus, Lawrence J., Charles J. Henny, David J. Hoffman, and Robert A. Grove. "Lead Toxicosis in Tundra Swans Near a Mining and Smelting Complex in Northern Idaho." *Archives of Environmental Contamination and Toxicology* 21, no. 4 (1991): 549–55.

Bocking, Stephen. *Ecologists and Environmental Politics: A History of Contemporary Ecology*. New Haven, CT: Yale University Press, 1997.

Bocking, Stephen. "Ecosystems, Ecologists, and the Atom: Environmental Research at Oak Ridge National Laboratory." *Journal of the History of Biology* 28, no. 1 (1995): 1–47.

Bocking, Stephen. *Nature's Experts: Science, Politics, and the Environment*. New Brunswick, NJ: Rutgers University Press, 2004.

Bryson, Jeremy. "The Nature of Gentrification: Urban Environments and Redevelopment in the Inland Northwest." PhD diss., Syracuse University, 2010.

Bullard, Robert D., ed. *Confronting Environmental Racism: Voices from the Grassroots*. Boston: South End Press, 1993.

Bullard, Robert D. *Dumping in Dixie: Race, Class, and Environmental Quality*. Boulder, CO: Westview Press, 1990.

"Bunker Hill & Sullivan Smelting Works at Kellogg, Idaho." *Engineering and Mining Journal*, May 13, 1916, 868–69.

Byrkit, James W. *Forging the Copper Collar: Arizona's Labor Management War of 1901–1921*. Tucson: University of Arizona Press, 1982.

Cantrell, John. "The Controversial Closure of the Bunker Hill Company—Revisiting the Question of Culpability." Master's thesis, Boise State University, 1996.

Casner, Nicholas A. "Leaded Waters: A History of Mining Pollution on the Coeur d'Alene River in Idaho, 1900–1950." Master's thesis, Boise State University, 1989.

Casner, Nicholas A. "Toxic River: Politics and Coeur d'Alene Mining Pollution in the 1930s." *Idaho Yesterdays*, Fall 1991, 2–19.

Chapman, Ray. *History of Idaho's Silver Valley: 1878–2000*. Kellogg, ID: Chapman, 2000.

Chapman, Ray. *History of Kellogg, Idaho: 1885–2002*. Kellogg, ID: Chapman, 2002.

Chapman, Ray. *Uncle Bunker: Memories in Words and Pictures*. Kellogg, ID: Chapman, 1994.

Chavkin, Wendy, ed. *Double Exposure: Women's Health Hazards on the Job and at Home*. New York: Monthly Review Press, 1984.

Chupp, Norman R., and Paul D. Dalke. "Waterfowl Mortality in the Coeur d'Alene River Valley, Idaho." *Journal of Wildlife Management* 28, no. 4 (1964): 692–702.

Cioc, Marc. *The Rhine: An Eco-Biography, 1815–2000*. Seattle: University of Washington Press, 2002.

Clark, Norman H. *Mill Town: A Social History of Everett, Washington*. Seattle: University of Washington Press, 1970.

Coeur d'Alene River and Lake Commission. *Kellogg Revisited—1983 [sic] Childhood Blood Lead and Environmental Status Report* [draft]. Idaho Department of Health and Welfare, 1984.

Coeur d'Alene River and Lake Commission. *Report and Recommendations of the Coeur d'Alene River and Lake Commission to the Twenty-Second Session of the State Legislature of Idaho*, 1933.

Cohen, Lizabeth. *A Consumers' Republic: The Politics of Mass Consumption in Postwar America*. New York: Knopf, 2003.

Coleman, Annie Gilbert. *Ski Style: Sport and Culture in the Rockies*. Lawrence: University Press of Kansas, 2004.

Cronon, William. "Kennecott Journey: The Paths Out of Town." In *Under an Open Sky: Rethinking America's Western Past*, ed. William Cronon, George Miles, and Jay Gitlin, 28–51. New York: W. W. Norton, 1992.

Cronon, William. *Nature's Metropolis: Chicago and the Great West*. New York: W. W. Norton, 1991.

Culver, Lawrence. "From 'Last of the Old West' to First of the New West: Tourism and Transformation in Jackson Hole, Wyoming." In Nicholas, Bapis, and Harvey, *Imagining the Big Open*, 163–80.

Dann, Kevin, and Gregg Mitman. "Exploring the Borders of Environmental History and the History of Ecology." Review of *Uncommon Ground: Toward Reinventing Nature*, by William Cronon. *Journal of the History of Biology* 30, no. 2 (1997): 291–302.

Davis, Mike. *City of Quartz: Excavating the Future in Los Angeles*. London: Verso, 1990.

Davis, Mike. *Ecology of Fear: Los Angeles and the Imagination of Disaster*. New York: Vintage, 1999.

DeBuys, William. *Enchantment and Exploitation: The Life and Hard Times of a New Mexico Mountain Range*. Albuquerque: University of New Mexico Press, 1985.

Demeritt, David. "Ecology, Objectivity, and Critique in Writings on Nature and Human Societies." *Journal of Historical Geography* 20, no. 1 (1994): 22–37.

Denworth, Lydia. *Toxic Truth: A Scientist, a Doctor, and the Battle over Lead*. Boston: Beacon, 2008.

Deverell, William, ed. *A Companion to the American West*. Malden, MA: Blackwell, 2004.

Deverell, William. "Politics and the Twentieth-Century American West." In Deverell, *Companion to the American West*, 442–59.

Edna Grace Yoss, et al. v. The Bunker Hill Company, et al., in the US District Court for the District of Idaho. (1977–1981).

Eichstaedt, Peter H. *If You Poison Us: Uranium and Native Americans*. Santa Fe: Red Crane, 1994.

Ellis, M. M., for Coeur d'Alene River and Lake Commission. *Special Scientific Report, No. 1, Pollution of the Coeur d'Alene River and Adjacent Waters by Mine Wastes*. Washington, DC: Department of the Interior, Bureau of Fisheries, 1940.

Forestell, Nancy. "'And I Feel Like I'm Dying from Mining for Gold': Disability, Gender, and the Mining Community, 1920–1950." *Labor: Studies in Working-Class History of the Americas* 3, no. 3 (2006): 77–93.

Francaviglia, Richard V. *Hard Places: Reading the Landscape of America's Historic Mining Districts*. Iowa City: University of Iowa Press, 1991.

Frey, Rodney, with the Schitsu'umsh. *Landscape Traveled by Coyote and Crane: The World of the Schitsu'umsh (Coeur d'Alene Indians)*. Seattle: University of Washington Press, 2001.

Gandy, Matthew. *Concrete and Clay: Reworking Nature in New York City*. Cambridge: MIT Press, 2002.

Garner, John S., ed. *The Company Town: Architecture and Society in the Early Industrial Age*. New York: Oxford University Press, 1992.

Hall, Jacquelyn Dowd, James Leloudis, Robert Korstad, Mary Murphy, Lu Ann Jones, and Christopher B. Daly. *Like a Family: The Making of a Southern Cotton Mill World*. Chapel Hill: University of North Carolina Press, 1987.

Hart, Patricia, and Ivar Nelson. *Mining Town: The Photographic Record of T. N. Barnard and Nellie Stockbridge from the Coeur d'Alenes*. Seattle: University of Washington Press, 1984.

Hays, Samuel P. *Beauty, Health, and Permanence: Environmental Politics in the United States, 1955–1985*. Cambridge, UK: Cambridge University Press, 1987.

Historic Wallace Preservation Society. *Images of America: The Silver Valley*. Charleston, SC: Arcadia, 2010.

Hofman, H. O. "Metallurgy of Lead in 1916." *Engineering and Mining Journal*, January 6, 1917, 49–51.

Hult, Ruby El. *Steamboats in the Timber*. Caldwell, ID: Caxton, 1952.

Hurley, Andrew. *Environmental Inequalities: Class, Race, and Industrial Pollution in Gary, Indiana, 1945–1980*. Chapel Hill: University of North Carolina Press, 1995.

Hyde, Anne. "Transients and Stickers: The Problem of Community in the American West." In Deverell, *Companion to the American West*, 304–28.

Hyde, Gene. *From Hell to Heaven: Death-Related Mining Accidents in North Idaho*. Coeur d'Alene, ID: Museum of North Idaho, 2003.

Idaho Secretary of State. *General Laws of the State of Idaho Passed at the Twenty-First Session of the State Legislature 1931*. Boise, ID: Capitol News Publishing, 1931.

Igler, David. "Engineering the Elephant: Industrialism and the Environment in the Greater West." In Deverell, *Companion to the American West*, 93–111.

Isenberg, Andrew C. "Environment and the Nineteenth-Century West: Or, Process Encounters Place." In Deverell, *Companion to the American West*, 77–92.

Isenberg, Andrew C. *Mining California: An Ecological History*. New York: Hill and Wang, 2005.

Jaffe, James Alan. *The Struggle for Market Power: Industrial Relations in the British Coal Industry, 1800–1840*. Cambridge, UK: Cambridge University Press, 1991.

Jensen, Dwight. "The Loneliness of the Environmental Reporter." *Columbia Journalism Review* 15, no. 5 (1977): 40–42.

Jessee, Emory Jerry. "Radiation Ecologies: Bombs, Bodies, and Environment during the Atmospheric Nuclear Weapons Testing Period in the United States, 1942–1965." PhD diss., Montana State University–Bozeman, 2013.

Johnston, Ronnie, and Arthur McIvor. "Oral History, Subjectivity, and Environmental Reality: Occupational Health Histories in Twentieth-Century Scotland." *Osiris* 19 (2004): 234–49.

Kemmerer, George, J. F. Bovard, and W. R. Boorman. *Northwestern Lakes of the United States: Biological and Chemical Studies with Reference to Possibilities of Production of Fish*. Bulletin of the Bureau of Fisheries 34. Washington, DC: Government Printing Office, 1923.

Kirsch, Scott. "Harold Knapp and the Geography of Normal Controversy: Radioiodine in the Historical Environment." *Osiris* 19 (2004): 167–81.

Kitman, Jamie Lincoln. "The Secret History of Lead." *Nation*, March 20, 2000.

Klingle, Matthew. *Emerald City: An Environmental History of Seattle*. New Haven, CT: Yale University Press, 2007.

Kuhn, Thomas S. *The Structure of Scientific Revolutions*. Chicago: University of Chicago Press, 1962.

Landrigan, Philip J., Edward L. Baker, Robert G. Feldman, Dennis H. Cox, Kenneth V. Eden, Walter A. Orenstein, John A. Mather, Anthony J. Yankel, and Ian von Lindern. "Increased Absorption with Anemia and Slowed Nerve Conduction in Children Near a Lead Smelter." *Journal of Pediatrics* 89, no. 6 (Dec. 1976): 904–10.

Landrigan, Philip J., Steven H. Gehlbach, Bernard F. Rosenblum, Jimmie M. Shoults, P. E. Robert, M. Candelaria, William F. Barthel, John A. Liddle, Anne L. Smrek, Norma W. Staehling, and JoDean F. Sanders. "Epidemic Lead Absorption Near an Ore Smelter—The Role of Particulate Lead." *New England Journal of Medicine* 292, no. 3 (Jan. 16, 1975): 123–29.

Langston, Nancy. *Where Land and Water Meet: A Western Landscape Transformed.* Seattle: University of Washington Press, 2003.

Latour, Bruno. *Science in Action: How to Follow Scientists and Engineers through Society.* Cambridge, MA: Harvard University Press, 1987.

Lavender, David, and David G. Lavender. *The Telluride Story: A Tale of Two Towns.* Ridgway, CO: Wayfinder, 2007.

LeCain, Timothy J. *Mass Destruction: The Men and Giant Mines That Wired America and Scarred the Planet.* New Brunswick, NJ: Rutgers University Press, 2009.

Lewty, Peter J. *Across the Columbia Plain: Railroad Expansion in the Interior Northwest, 1885–1893.* Pullman: Washington State University Press, 1995.

Long, Keith R. "Tailings under the Bridge: Causes and Consequences of River Disposal of Tailings, Coeur d'Alene Mining Region, 1886 to 1968." *Mining History Journal* 8 (2001): 83–101.

Lukas, J. Anthony. *Big Trouble: A Murder in a Small Western Town Sets off a Struggle for the Soul of America.* New York: Simon and Schuster, 1997.

Lupton, Deborah. *Risk.* London: Routledge, 1999.

Lutts, Ralph H. "Chemical Fallout: Rachel Carson's *Silent Spring,* Radioactive Fallout, and the Environmental Movement." *Environmental Review* 9, no. 3 (1985): 210–25.

MacMillan, Donald. *Smoke Wars: Anaconda Copper, Montana Air Pollution, and the Courts, 1890–1924.* Helena: Montana Historical Society Press, 2000.

Marcosson, Isaac. *Metal Magic: The Story of the American Smelting and Refining Company.* New York: Farrar, Strauss, 1949.

Markowitz, Gerald, and David Rosner. *Deceit and Denial: The Deadly Politics of Industrial Pollution.* Berkeley: University of California Press, 2002.

Marquardt, Steve. "Pesticides, Parakeets, and Unions in the Costa Rican Banana Industry, 1938–1962." *Latin American Research Review* 37, no. 2 (2002): 3–36.

McEvoy, Arthur F. *The Fisherman's Problem: Ecology and Law in the California Fisheries, 1850–1980.* Cambridge, UK: Cambridge University Press, 1986.

McEvoy, Arthur F. "Working Environments: An Ecological Approach to Industrial Health and Safety." In "Snapshots of a Discipline: Selected Proceedings from the Conference on Critical Problems and Research Frontiers in the History of Technology, Madison, Wisconsin, October 30–November 3, 1991." Supplement, *Technology and Culture* 36, no. 2 (1995): S145–73.

McFarlane, Wallace Scot. "Defining a Nuisance: Pollution, Science, and Environmental Politics on Maine's Androscoggin River." *Environmental History* 17, no. 2 (2012): 307–35.

McGurty, Eileen Maura. "From NIMBY to Civil Rights: The Origins of the Environmental Justice Movement." *Environmental History* 2, no. 3 (1997): 301–23.

Mercier, Laurie. *Anaconda: Labor, Community, and Culture in Montana's Smelter City.* Urbana: University of Illinois Press, 2001.

Mercier, Laurie. "Borders, Gender, and Labor: Canadian and U.S. Mining Towns during the Cold War Era." In *Company Towns in the Americas: Landscape, Power, and Working-Class Communities,* ed. Oliver J. Dinius and Angela Vergara, 158–77. Athens: University of Georgia Press, 2011.

Michiko, Ishimure. *Paradise in the Sea of Sorrow: Our Minamata Disease.* Ann Arbor: Center for Japanese Studies at the University of Michigan, 1990.

Mitchell, Don. "Controlling Space, Controlling Scale: Migratory Labour, Free Speech, and Regional Development in the American West." *Journal of Historical Geography* 28, no. 1 (2002): 63–84.

Mitman, Gregg. "Geographies of Hope: Mining the Frontiers of Health in Denver and Beyond, 1870–1965." *Osiris* 19 (2004): 93–111.

Mitman, Gregg. "In Search of Health: Landscape and Disease in American Environmental History." *Environmental History* 10, no. 2 (2005): 184–210.

Mitman, Gregg, Michelle Murphy, and Christopher Sellers. "Introduction: A Cloud over History." *Osiris* 19 (2004): 1–17.

Morse, Dale L., Philip J. Landrigan, Bernard F. Rosenblum, Jaqueline S. Hubert, and Jere Housworth. "El Paso Revisited: Epidemiologic Follow-up of an Environmental Lead Problem." *JAMA* 242, no. 8 (1979): 739–41.

Murphy, Mary. *Mining Cultures: Men, Women, and Leisure in Butte, 1914–41*. Urbana: University of Illinois Press, 1997.

Murphy, Michelle. "Uncertain Exposures and the Privilege of Imperception: Activist Scientists and Race at the U.S. Environmental Protection Agency." *Osiris* 19 (2004): 266–82.

Nash, Linda. "Finishing Nature: Harmonizing Bodies and Environments in Late-Nineteenth-Century California." *Environmental History* 8, no. 1 (2003): 25–52.

Nash, Linda. "The Fruits of Ill-Health: Pesticides and Workers' Bodies in Post–World War II California." *Osiris* 19 (2004): 203–19.

Nash, Linda. *Inescapable Ecologies: A History of Environment, Disease, and Knowledge*. Berkeley: University of California Press, 2006.

National Research Council, Committee on Superfund Site Assessment in the Coeur d'Alene River Basin. *Superfund and Mining Megasites: Lessons from the Coeur d'Alene River Basin*. Washington, DC: National Academies Press, 2005.

Needleman, Herbert L., Charles Gunnoe, Alan Leviton, Robert Reed, Henry Peresie, Cornelius Maher, and Peter Barrett. "Deficits in Psychologic and Classroom Performance of Children with Elevated Dentine Lead Levels." *New England Journal of Medicine* 300, no. 13 (March 29, 1979): 689–95.

Nelson, Daniel. "The Company Union Movement, 1900–1937: A Reexamination." *Business History Review* 56, no. 3 (1982): 335–57.

Nicholas, Liza, Elaine M. Bapis, and Thomas J. Harvey, eds. *Imagining the Big Open: Nature, Identity and Play in the New West*. Salt Lake City: University of Utah Press, 2003.

Olsen, Gregg. *The Deep Dark: Disaster and Redemption in America's Richest Silver Mine*. New York: Three Rivers, 2005.

Palmer, Tim. *America by Rivers*. Washington, DC: Island, 1996.

Parmalee, H. C. "The Bunker Hill and Sullivan Enterprise Today." *Engineering and Mining Journal* 140 (Aug. 1939): 35–83.

Peacock, Andrea. *Libby, Montana: Asbestos and the Deadly Silence of an American Corporation*. Boulder, CO: Johnson, 2003.

Perales, Monica. "Fighting to Stay in Smeltertown: Lead Contamination and Environmental Justice in a Mexican American Community." *Western Historical Quarterly* 39, no. 1 (2008): 41–63.

Perales, Monica. *Smeltertown: Making and Remembering a Southwest Border Community*. Chapel Hill: University of North Carolina Press, 2010.

Petersen, Keith C. *Company Town: Potlatch, Idaho, and the Potlatch Lumber Company*. Pullman: Washington State University Press, 1987.

Petryna, Adriana. "Biological Citizenship: The Science and Politics of Chernobyl-Exposed Populations." *Osiris* 19 (2004): 250–65.

Pritchard, Sara. *Confluence: The Nature of Technology and the Remaking of the Rhône*. Cambridge, MA: Harvard University Press, 2011.

Quivik, Fredric. "Smoke and Tailings: An Environmental History of Copper Smelting Technologies in Montana, 1880–1930." PhD diss., University of Pennsylvania, 1998.

Quivik, Frederik L. "Of Tailings, Superfund Litigation, and Historians as Experts: *U.S. v Asarco, et al.* (the Bunker Hill Case in Idaho)." *Public Historian* 26, no. 1 (2004): 81–104.

Rabe, Fred W., and David C. Flaherty. *The River of Green and Gold: A Pristine Wilderness Dramatically Affected by Man's Discovery of Gold*. Moscow, ID: Idaho Research Foundation, 1974.

Ragaini, R. C., H. R. Ralston, and N. Roberts. "Environmental Trace Metal Contamination in Kellogg, Idaho, near a Lead Smelting Complex." *Environmental Science and Technology* 11, no. 8 (Aug. 1977): 773–81.

Riebsame, William, ed. *Atlas of the New West: Portrait of a Changing Region*. New York: W. W. Norton, 1997.

Robbins, Paul. *Political Ecology: A Critical Introduction*. Malden, MA: Blackwell, 2004.

Robbins, William G. *Colony and Empire: The Capitalist Transformation of the American West*. Lawrence: University Press of Kansas, 1994.

Robbins, William G. *Hard Times in Paradise: Coos Bay, Oregon, 1850–1986*. Seattle: University of Washington Press, 1988.

Robbins, William G. *Lumberjacks and Legislators: Political Economy of the U.S. Lumber Industry, 1890–1941*. College Station: Texas A&M University Press, 1982.

Rosner, David, and Gerald Markowitz, eds. *Dying for Work: Workers' Safety and Health in Twentieth-Century America*. Bloomington: Indiana University Press, 1987.

Rothman, Hal K. *Devil's Bargains: Tourism in the Twentieth-Century American West*. Lawrence: University Press of Kansas, 1998.

Seager, Henry R. "Company Unions vs. Trade Unions." *American Economic Review* 13, no. 1 (Mar. 1923): 1–13.

Selevan, S. G., P. J. Landrigan, F. B. Stern, and J. H. Jones. "Mortality of Lead Smelter Workers." *American Journal of Epidemiology* 122, no. 4 (1985): 673–83.

Sellers, Christopher. "The Artificial Nature of Fluoridated Water: Between Nations, Knowledge, and Material Flows." *Osiris* 19 (2004): 182–200.

Sellers, Christopher. "Body, Place and the State: The Makings of an 'Environmentalist' Imaginary in the Post–World War II U.S." *Radical History Review* 74 (1999): 31–64.

Sellers, Christopher. "Factory as Environment: Industrial Hygiene, Professional Collaboration and the Modern Sciences of Pollution." *Environmental History Review* 18, no. 1 (1994): 55–83.

Sellers, Christopher. *Hazards of the Job: From Industrial Disease to Environmental Health Science*. Chapel Hill: University of North Carolina Press, 1997.

Sellers, Christopher. "Thoreau's Body: Towards an Embodied Environmental History." *Environmental History* 4, no. 4 (1999): 486–514.

Schwantes, Carlos A. *In Mountain Shadows: A History of Idaho*. Lincoln: University of Nebraska Press, 1991.

Scott, James C. *Seeing Like a State: How Certain Schemes to Improve the Human Condition Have Failed*. New Haven, CT: Yale University Press, 1998.

Shaffer, Marguerite S. "'The West Plays West': Western Tourism and the Landscape of Leisure." In Deverell, *Companion to the American West*, 375–89.

Smith, Duane A. *Mining America: The Industry and the Environment, 1800–1980*. Lawrence: University Press of Kansas, 1987.

Stapilus, Randy. *1990: The Idaho Political Almanac*. Boise, ID: Ridenbaugh, 1990.

Stradling, David. *Smokestacks and Progressives: Environmentalists, Engineers, and Air Quality in America, 1881–1951*. Baltimore: Johns Hopkins University Press, 1999.

Tarr, Joel. *The Search for the Ultimate Sink: Urban Pollution in Historical Perspective*. Akron, OH: University of Akron Press, 1996.

Tate, Cassandra. "American Dilemma of Jobs, Health in an Idaho Town." *Smithsonian* 12 (Sept. 1981): 74–83.

Tone, Andrea. *The Business of Benevolence: Industrial Paternalism in Progressive America*. Ithaca, NY: Cornell University Press, 1997.

Travis, William R. *New Geographies of the American West: Land Use and the Changing Patterns of Place*. Washington, DC: Island, 2007.

Troesken, Werner. *The Great Lead Water Pipe Disaster*. Cambridge, MA: MIT Press, 2006.

US Public Health Service, Division of Water Pollution Control. *Pacific Northwest Drainage Basins*. Washington, DC: Government Printing Office, 1951.

Valenčius, Conevery Bolton. *The Health of the Country: How American Settlers Understood Themselves and Their Land*. New York: Basic Books, 2002.

Walker, Brett L. *Toxic Archipelago: A History of Industrial Disease in Japan*. Seattle: University of Washington Press, 2010.

Ward, Chip. *Canaries on the Rim: Living Downwind in the West*. London: Verso, 1999.

Warren, Christian. *Brush with Death: A Social History of Lead Poisoning*. Baltimore: Johns Hopkins University Press, 2000.

Wegner, Glen, ed. *Shoshone Lead Health Project: Work Summary*. Boise: Idaho Department of Health and Welfare, 1976.

Weston, Julie Whitesel. *The Good Times Are All Gone Now: Life, Death, and Rebirth in an Idaho Mining Town*. Norman: University of Oklahoma Press, 2009.

White, Richard. *Land Use, Environment, and Social Change: The Shaping of Island County, Washington*. Seattle: University of Washington Press, 1980.

White, Richard. *The Organic Machine: The Remaking of the Columbia River*. New York: Hill and Wang, 1995.

Wiltsie, Meredith, and William Wyckoff. "Reinventing Red Lodge: The Making of a New Western Landscape, 1884–2000." In Nicholas, Bapis, and Harvey, *Imagining the Big Open*, 124–50.

Wirth, John D. *Smelter Smoke in North America: The Politics of Transborder Pollution*. Lawrence: University Press of Kansas, 2000.

Wirth, John D. "The Trail Smelter Dispute: Canadians and Americans Confront Transboundary Pollution, 1927–41." *Environmental History* 1, no. 2 (1996): 34–51.

Winter, Thomas. *Making Men, Making Class: The YMCA and Workingmen, 1877–1920*. Chicago: University of Chicago Press, 2002.

Wolff, David A. *Industrializing the Rockies: Growth, Competition and Turmoil in the Coalfields of Colorado and Wyoming, 1868–1914*. Boulder: University Press of Colorado, 2003.

Worster, Donald. *Rivers of Empire: Water, Aridity, and the Growth of the American West*. New York: Oxford University Press, 1992.

Young, Biloine Whiting. *River of Conflict, River of Dreams: Three Hundred Years on the Upper Mississippi*. Saint Paul, MN: Pogo, 2004.

INDEX